Geomechanics in CO_2 Storage Facilities

Geomechanics in CO_2 Storage Facilities

Edited by
Gilles Pijaudier-Cabot
Jean-Michel Pereira

ISTE

WILEY

First published 2013 in Great Britain and the United States by ISTE Ltd and John Wiley & Sons, Inc.

ISTE Ltd
27-37 St George's Road
London SW19 4EU
UK

www.iste.co.uk

John Wiley & Sons, Inc.
111 River Street
Hoboken, NJ 07030
USA

www.wiley.com

Library of Congress Control Number: 2012950755

British Library Cataloguing-in-Publication Data
A CIP record for this book is available from the British Library
ISBN: 978-1-84821-416-3

MIX
Paper from responsible sources
FSC
www.fsc.org
FSC® C013604

Table of Contents

Preface

Geological storage of CO_2 aims at injecting carbon dioxide (CO_2) into a geological formation capable of ensuring the durable storage of large quantities of this gas, and preventing (or at least minimizing) the harmful effects of CO_2 emissions and environmental and health risks. The development of this technology follows the Intergovernmental Panel on Climate Change (IPCC) recommendations to reduce anthropic greenhouse gas emissions into the atmosphere.

Apart from questions related to its economical feasibility, the development of this technology at an industrial scale relies on its societal acceptability. The importance of ensuring societal acceptance is reinforced by the fact that potential sites for geological storage of CO_2 are located in relatively densely inhabited areas, at least in Europe. From a scientific and technical point of view, this acceptability requires a sound and rigorous assessment of the potential risks associated with this geological storage. Geological storage of CO_2 has to indeed remain environmentally safe for relatively long periods of time – typically a thousand years, which actually is much longer than usual periods related to conventional civil or petroleum engineering applications.

Existing scientific studies of geological storage of CO_2 generally focus on geological or geochemical aspects such as those encountered in reactive transport studies at the material or reservoir scales. Recently, research efforts have been directed toward geomechanical issues related to the chemical reactivity of CO_2 with the geomaterials located within the storage site (well cement, reservoir rock, cap rock, etc.).

This book comprises contributions from leading scientific experts presented at the 22nd ALERT Workshop held in Aussois, France, in October 2011. ALERT Geomaterials is the acronym for the "Alliance of Laboratories in Europe for Research and Technology on geomaterials". This network gathers 27 universities or organizations, mostly in Europe, which are most active in the field of numerical and experimental modeling of geomaterials and geostructures (www.alertgeomaterials.eu).

During this workshop, a session dedicated to "Geomechanical issues of CO_2 storage" was organized by the coordinators of this volume, and it was decided to collect the presentations in order to provide scientists with a panorama of recent advances on this important environmental issue. This book collects 11 contributions, covering three main topics:

– The first topic concerns *transport processes*. Chapter 1 by Broseta and Chapter 2 by Gérard *et al.* deal with assessing steel rock integrity and gas transport in clayey materials, respectively, thus addressing the question of the hydraulic performance and tightness of the caprock. Chapter 3 by Khaddour *et al.* presents a methodology to estimate transport properties of geomaterials from their pore size distributions.

– The second topic focuses on *fracture, deformation and coupled effects*. Chapter 4 by Seyedi *et al.* and Chapter 5 by Selvadurai *et al.* address the question of the mechanical integrity of the caprock in the context of CO_2 storage. Chapter 6 by Nouailletas *et al.* studies the problem of fault reactivation due to its chemical degradation by CO_2. Chapter 7 by Nikoosokhan *et al.* concerns the particular case of CO_2 storage in coal seams and presents a framework encompassing adsorption effects on the poromechanical behavior of coal.

– The last topic includes four chapters dedicated to *ageing and integrity* issues. Chapter 8 by Lewandowska studies the geomechanical effects of rock dissolution induced by acid fluid attacks using up-scaling techniques. Chapter 9 by Shao *et al.*, Chapter 10 by Shen *et al.* and Chapter 11 by Pereira and Vallin focus on the behavior of well cements subjected to chemical alterations due to the presence of CO_2.

We would like to thank all of the authors of this book for their kind cooperation and timely contributions. We also extend our grateful thanks to ALERT Geomaterials and its past director, Professor Claudio di Prisco, for taking the initiative of setting this topic on the agenda of the annual workshop of this network, and thus providing the ground and logistics for lively and stimulating scientific exchanges, which were the origin of this book.

Gilles PIJAUDIER-CABOT
University of Pau and Pays de l'Adour

Jean-Michel PEREIRA
Ecole des Ponts ParisTech
November 2012

PART 1

Transport Processes

Chapter 1

Assessing Seal Rock Integrity for CO_2 Geological Storage Purposes

1.1. Introduction

Prior to any CO_2 geological storage project, an assessment is needed to confirm the ability of the formations overlying the storage reservoir to act as a seal with respect to the underlying pressurized CO_2. In the case of hydrocarbon reservoirs, these formations typically consist of low-permeability caprocks: the ability of these tight rocks (usually imbibed with water) to act as a seal with respect to an underlying buoyant phase is primarily a capillary effect, which is indeed proven in the case of the hydrocarbon phase present in the reservoir, but not in the case of a CO_2-rich phase. In the case of deep saline aquifers, the sealing efficiency of the overlying tight formations (referred to as aquitards or aquicludes) with respect to a non-aqueous buoyant fluid is not known.

Excessive pressurization of the CO_2 phase, due to the injection operations and to the buoyancy effect, induces leakage through the seal rock(s) by a variety of mechanisms, each being active above a specific pressure threshold. The lowest of these thresholds should indeed not be exceeded during the CO_2 injection and storage operations – unless the associated leakage rate is extremely small.

The failure mechanisms are conventionally divided into two types: mechanical and capillary. Seals that fail because of mechanical and capillary mechanisms are referred to as hydraulic seals and membrane seals, respectively [WAT 87].

Chapter written by Daniel BROSETA.

Mechanical failure comprises the development of high-permeability pathways through a variety of processes, such as tensile fracturing or shear slip reactivation of pre-existing faults, as well as thermal fracturing. This is considered to be the most dangerous failure mode, with high associated leakage rates, and it is constantly monitored in most ongoing CO_2 storage projects, e.g. by passive microseismic techniques (see [RUT 12] for a recent review).

Capillary failure (or capillary invasion), in which the rock pore structure is left unaltered during (and after) the gas leakage process, is a more pervasive process than mechanical failure, and lower leakage rates are expected – the effective gas permeability that follows gas capillary breakthrough is only a fraction of the intrinsic permeability of the seal rock, which itself does (or should) not exceed a few micro-Darcy (10^{-18} m^2). It is expected to occur at a lower CO_2 (pore) pressure than the pressure required for tensile fracturing, except in extremely tight rocks such as evaporites [GLU 04]. This pressure corresponds to the capillary entry pressure, or the minimum overpressure in the CO_2 phase (compared to brine pressure) required to intrude into the seal rock and to displace the saturating brine phase. From the Laplace–Young equation, this overpressure, hereafter denoted by P_c, is inversely proportional to a characteristic pore throat size R and proportional to the interfacial tension (IFT) between the brine and the stored CO_2-rich gas ($\gamma_{w,g}$) and to the cosine of the contact angle θ characterizing the wettability of the seal rock:

$$P_c \sim \gamma_{w,g} \cos(\theta)/R \qquad [1.1]$$

More precisely, θ is the water-receding (or gas-advancing) angle on the rock substrate, which is referred to in the following as the drainage angle. This angle is often significantly smaller than the imbibition angle corresponding to the reverse displacement (water displacing the CO_2-rich fluid from the substrate): the difference, or contact angle hysteresis, is related in a complex manner to the interactions of CO_2 with the substrate, and to substrate chemical and/or geometrical heterogeneities.

In practice, the sealing capacity of a tight rock (sampled for instance from the overburden of a potential storage site) initially saturated with brine is appreciated in the laboratory from breakthrough or displacement experiments. This sort of experiment is carried out in a more general context, e.g. for qualifying the confining properties of the engineered (compact) clay barriers used for underground nuclear waste disposal (here, the gas is the hydrogen generated by the waste) [GAL 98, HAR 99, HOR 99, GAL 00], or of the upper layers of underground reservoirs for seasonal storage of natural gas. The "standard" procedure [THO 68] for running such experiments consists of increasing stepwise the gas pressure on one face of a brine-saturated rock sample until the gas starts penetrating into the sample and thus displaces the saturating brine, which defines the gas entry pressure, and then, for a slightly larger pressure threshold called breakthrough or displacement pressure, comes out on the other face of the sample; then, the effective gas permeability after breakthrough can also be determined. The gas (CO_2) breakthrough (or displacement) pressure, hereafter denoted as P_d, and the effective gas permeability after breakthrough are very important in practice, as they are related, respectively, to the maximum sustainable overpressure in the storage reservoir and to the leakage rate of the CO_2-rich fluid following breakthrough (P_d can be exceeded, however, if this rate is extremely small). If it is assumed that the capillary invasion process described in the previous paragraph is the dominant failure mechanism (i.e. the seal rock is a membrane seal), then P_d should obey the following Laplace–Young equation:

$$P_d - P_w \sim \gamma_{w,g} \cos(\theta)/r \sim \gamma_{w,g} \cos(\theta)/k^{1/2} \qquad [1.2]$$

where P_w is the pressure in the brine that imbibes the seal rock, r is the size of the narrowest pore throats along the leakage pathway in the seal rock, and $k < r^{1/2}$ is the intrinsic (or absolute) permeability of the seal rock. In addition to being very tight (r and k small), the interfacial tension between brine and the CO_2-rich gas must be high enough and the seal rock must be water-wet (i.e. small θ) if the brine-saturated rock is to function as an effective (membrane) seal with respect to the buoyant CO_2-rich phase. On the one hand, the available evidence is that IFTs between brine and CO_2 in geological storage conditions are higher than ~23 mN/m – this value has to be corrected for the effects of the impurities co-injected with CO_2 and of the mixing with indigenous hydrocarbons (if storage is in a hydrocarbon reservoir). On the other hand, the water-wet character of many rocks and rock-bearing minerals seems to be preserved in the presence of CO_2, at least in the drainage conditions (CO_2 displaces water from the substrate) representative of the leakage process. This subject is currently being intensely investigated.

The outline of this chapter is as follows: the "standard" procedure for determining gas breakthrough (or displacement) pressures in water-saturated tight rocks and the ensuing gas effective permeability is presented in section 1.2. This determination, which is extremely time-consuming, is mandatory in the case of deep aquifers where there is no *a priori* knowledge of the sealing behavior of the top aquitard (or aquiclude), but it can be skipped in the case of hydrocarbon reservoirs where a lower bound on breakthrough pressure can be simply inferred from the reservoir discovery pressure and the relevant interfacial properties. These properties, namely the IFTs between brine and CO_2-rich fluids and the wetting (contact) angle θ, are reviewed and discussed in section 1.3, and the method for inferring a lower bound on the CO_2 breakthrough pressure of depleted hydrocarbon reservoirs is presented in section 1.4. In section 1.5, the gas breakthrough experiments available in the literature are further examined and discussed. It turns out that the mechanism of capillary invasion in a rigid porous medium (i.e. membrane seal failure) does not account for many of the observed features, which witness some quasi-permanent deformations of the porous structure.

The important topics of mechanical failure processes such as thermal fracturing, reconnection of fracture networks, and shear reactivation of faults (especially when a reactive fluid, i.e. CO_2 and water, changes the slip conditions) are not addressed in this chapter (see [RUT 12] for a review).

1.2. Gas breakthrough experiments in water-saturated rocks

The standard procedure [THO 68] for determining the gas breakthrough (or displacement) pressure in a given rock sample fully saturated with brine and placed in representative temperature and confining pressure conditions consists of increasing the gas pressure in a stepwise fashion (each step lasting a few hours to a few days) on one face of the sample until the breakthrough pressure is reached, i.e. gas displaces the saturating brine phase across the sample [LI 05, HAR 09, TON 10, TON 11, BOU 11, SKU 12]. Gas entry (but no breakthrough) may occur for the pressure step previous to breakthrough, i.e. the gas phase enters the rock, but does not come out on the other face of the sample. Most gas displacement experiments are conducted with nitrogen as the gas phase rather than with CO_2: then, the CO_2-breakthrough overpressure is obtained by multiplying the nitrogen-breakthrough overpressure by the ratio of the brine-CO_2 IFT to the brine-nitrogen IFT (assuming contact angles to be the same for the two gases: this point is discussed at the end of the following section).

In practice, the cylindrical (thin) rock sample to be tested is placed in a Hassler-type cell and fully saturated with brine under a prescribed representative confining pressure (this confining pressure or, more precisely, the effective pressure, i.e. the difference between the confining and pore pressures, has a strong impact on the rock petro-physical properties of interest; see [TON 10, TON 11, SKU 12]). The intrinsic permeability can then be determined, e.g. from steady-state measurements of the brine flow rate versus pressure drop across the sample [BOU 12]. Then, the gas phase is brought into contact with the inlet face of the sample and compressed in a stepwise manner, while a constant backpressure P_w is imposed in the brine phase on the outlet face, e.g. by means of a syringe pump working in the constant-pressure mode (the flow line from the sample outlet to the syringe is filled with the brine phase). Figure 1.1 displays the volume readings of the downstream pump and the upstream pressure in the CO_2 phase obtained in the course of a breakthrough (or displacement) experiment. The sample in this example comes from the carbonate-rich overburden of a depleted gas reservoir currently used as a CO_2 storage reservoir in the southwest of France. The absence of brine flow (as indicated by the volume readings in the downstream pump) indicates that CO_2 breakthrough has not yet occurred for the maximum investigated CO_2 (upstream) pressure of P_g = 8.6 MPa (the experiment was stopped at that pressure), which is 7.6 MPa above the pore (brine) pressure P_w = 1 MPa (see [TON 10, TON 11] for details): the CO_2 entry and breakthrough overpressures are thus higher than 7.6 MPa.

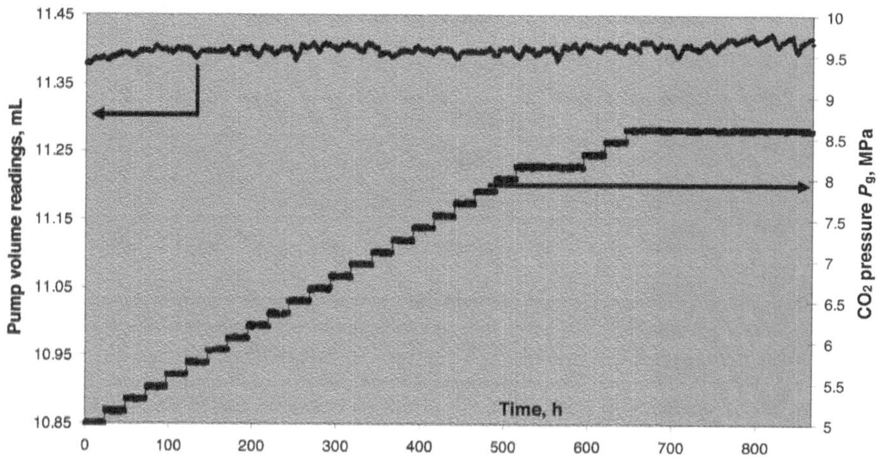

Figure 1.1. *In this experiment, breakthrough has not yet occurred for gas (upstream) pressures as high as 8.6 MPa. Rock permeability is in the nano-Darcy range, and the brine pressure downstream is equal to 1 MPa [TON 10, TON 11]*

Figure 1.2 displays the results obtained in a similar experiment carried out with a more permeable caprock sample, in which CO_2 entry (but not breakthrough) is apparent for the first pressure step investigated (equal to 1.5 MPa, i.e. 0.5 MPa above the brine pressure), whereas CO_2 flows continuously through the sample for the following pressure step P_g = 2.5 MPa, as manifested by the uninterrupted displacement of the piston of the downward pump. The displacement pressure P_d is thus intermediate between 0.5 and 1.5 MPa. The gas effective permeability for the subsequent pressure step can be inferred from the measured steady flow rate Q and the equation

$$k_g = 2\ \mu_g\, QLP_w / A(P_g^2 - P_w^2) \qquad [1.3]$$

where L is the sample thickness and μ_g the viscosity of the gas (assumed to be a perfect gas). In this experiment, $k_g \approx$ 1.5 nano-Darcy.

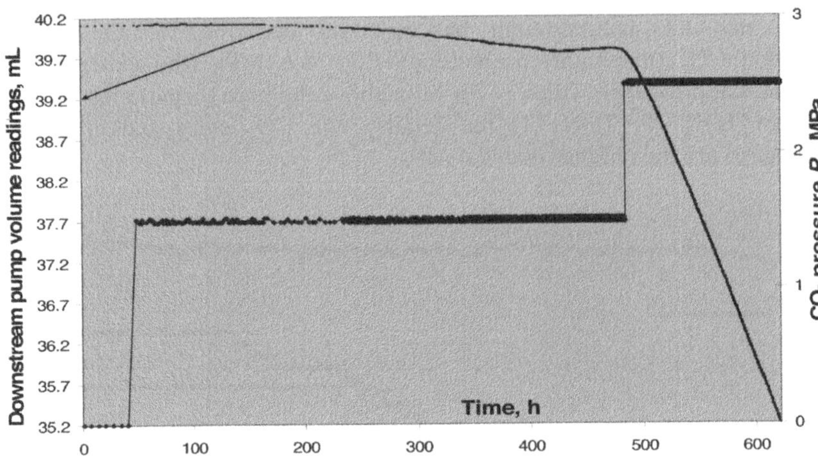

Figure 1.2. *In this experiment, carried out with the same setup as that used in Figure 1.1 and a more permeable rock sample, CO_2 entry into the rock sample is apparent when P_g = 1.5 MPa (first pressure step), and CO_2 breakthrough pressure is intermediate between P_g = 1.5 MPa and P_g =2.5 MPa, where gas flows through the rock sample at a steady rate, from which an effective gas permeability of 1.5 nano-Darcy is inferred (see text, and [TON 10, TON 11])*

The accuracy of the standard procedure is directly related to the amplitude and duration of the pressure steps and to the ability of the downstream syringe pump to record small volume variations or piston displacements. In the experiment of Figure 1.2, gas entry is apparent only 200–300 hours after the upstream pressure has been raised to 1.5 MPa, which suggests that in the experiment of Figure 1.1 all pressure steps except the last step were too short. An experiment typically takes weeks or even months; more than a year has been necessary to determine the consolidation behavior and gas (nitrogen) breakthrough parameters of a specimen of Nordland shale overlying the Sleipner CO_2 storage site [HAR 09].

The time required for accurate measurements is longer when the rock is less permeable. Experimental time can be saved, however, if mercury intrusion capillary pressure (MICP) data are available for the rock investigated (the reliability of such data is questionable for very fine-grained sediments [HIL 03, NEW 04]: they can only be used to give an order of magnitude). MICP measurements consist of forcing mercury (Hg) into an evacuated (dried) core sample under increasing pressure while monitoring the volume of Hg (the non-wetting phase) that is injected at each pressure step (the wetting phase is the mercury vapor). An estimate of the mercury capillary entry pressure is obtained as the intersection of the low Hg-saturation ($S_{Hg} \approx 0$) and intermediate Hg-saturation parts of the logarithm of the capillary pressure versus Hg-saturation [CAR 10]: this entry pressure must then be converted to that of the brine/gas system interest by using the proper interfacial tensions and contact angles (usually assumed to be equal to 140° in the mercury phase and to 0° in the brine).

A few alternate procedures have been proposed to evaluate breakthrough pressures in a shorter amount of time; they have recently been discussed and compared by Boulin *et al.* [BOU 11], who concluded that the standard procedure is the most representative of the leakage process of interest, even though duration is its main drawback. One of these alternate procedures [HIL 02, HIL 04] consists of forcing the gas into the brine-saturated rock at a pressure much higher than the breakthrough pressure, and then letting the system relax until the pressures in the inlet and outlet faces are stable: the residual pressure difference is considered to be a proxy of the breakthrough overpressure. This procedure provides meaningful trends but, as argued by Zweigel *et al.* [ZWE 04], the residual pressure difference (also called "snap-off pressure" [AMM 12]) results from a reimbibition process and is therefore expected to be smaller than the true breakthrough (or displacement) overpressure corresponding to a (first) drainage process. A factor of ~2 between the two quantities has in fact been measured by Harrington *et al.* [HAR 09] with the Nordland shale mentioned above, by Boulin *et al.* [BOU 11] with two other rock samples (one being the caprock of the Ketzin storage site in Germany), and by Amman-Hildenbrand *et al.* [AMM 12] as well.

1.3. Interfacial properties involved in seal rock integrity

As emphasized above, a key role is played by the brine/gas IFT and the wetting (or contact) angle θ, which have been thoroughly investigated over the past decade.

1.3.1. *Brine-gas IFT*

A large amount of brine/gas IFT values are available in the literature, which also provides simple methods for evaluating this property, including when the gas not only contain CO_2 but also some other "impurities" (such as H_2S, CH_4, N_2, etc.). One of these methods has recently been proposed by Duchateau and Broseta [DUC 12] and is detailed below.

Early measurements of IFT between pure water and CO_2 in pressure and temperature conditions relevant to geological storage have been critically reviewed by Georgiadis *et al.* [GEO 10] and Bikkina *et al.* [BIK 11a], who also presented some new measurements. Starting at low pressures from values close to the surface tension of pure water at the temperature of interest, the IFT between pure water and CO_2 decreases with increasing pressure, first linearly (for P below 3–5 MPa), then more slowly until it reaches a pseudo-plateau of about 23–30 mN/m for high-enough

P (larger than about 15–20 MPa at $T = 100°C$, and 8–10 MPa at $T = 40°C$), with the lowest values obtained after long (~1 day) equilibration times [BIK 11a]; this pseudo-plateau value turns out to be fairly independent of temperature up to 110°C.

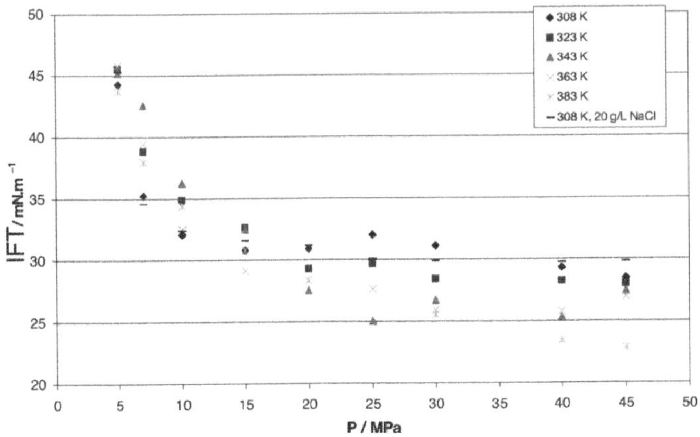

Figure 1.3. *Pressure dependence of the interfacial tension between CO_2 and pure water at $T = 35$, 50, 70, 90 and 100°C, and between CO_2 and a 20 g/L NaCl brine at 35°C [CHI 07a]*

When salts are added to the water, the IFT between the aqueous phase and the gas increases by an amount that is independent of pressure: this amount is thus equal to the increase in water surface tension (i.e. the IFT between the aqueous phase and its vapor or air). It can be rigorously shown from an analysis of Gibbs adsorption equation [DUC 12] that the effects of gas pressure (which tends to lower the IFT) and brine salinity (which tends to increase the IFT) on IFT are independent. Hence, the brine/gas IFT at given T and P is equal to the pure water/gas IFT at the same T and P, plus the surface tension increase induced by the addition of the brine salts to the water at T. This result holds for any gas that is sparingly soluble in water (including CO_2, and possibly H_2S) and for any non-adsorbing salt species [DUC 12]. The surface tension values of many salted aqueous solutions are available in the literature [ABR 93], as are the IFTs between pure water and CO_2 mixed with the following "impurities": CH_4 [REN 00] and N_2 [YAN 01], which tend to raise the IFT, and H_2S [SHA 08], which tends to lower the IFT. If one (or two) of these two tensions is (are) not available, it is easier to measure it (them) than the brine/gas IFT itself, as it does not require the use of corrosion-resistant equipment.

1.3.2. *Wetting behavior*

No clear-cut answer has yet been given as to whether or not rocks and rock-forming minerals remain water-wet in the presence of CO_2 in geological storage conditions, i.e. the contact angle θ in equations [1.1] and [1.2] remains low. Most of the recent experimental work in this domain consists either of contact angle measurements by the sessile drop method, or of direct observations of CO_2 and brine distributions in silica or glass micromodels. Using the sessile drop method, Yang *et al.* [YAN 08] observed that θ increases to values above 90° at high pressures in the CO_2 phase (>10 MPa) for a brine and a limestone rock from the Weyburn CO_2 storage reservoir; however, Espinosa and Santamarina [ESP 10] did not find any significant angle variation for quartz and calcite when increasing the pressure to 10 MPa at 23.5°C; Bikkina [BIK 11b] observed an increase in θ – by more than 40° – after repeated exposure of calcite and quartz to CO_2 at 25 and 50°C and pressures up to 20 MPa; Jung and Wan [JUN 12] observed for silica substrates at 45°C that θ increases with pressure by slightly less than 20°, much of this increase taking place between 7 and 10 MPa; more moderate variations have been recently observed for other water-wet substrates by Mills *et al.* [MIL 11] and by Wang *et al.* [WAN 12]. In direct observations in silica or glass micromodels, Chalbaud *et al.* [CHA 09] did not notice any significant change in wetting angles when displacing water with CO_2 under various subcritical (gaseous or liquid) and supercritical conditions (up to 60°C and 10 MPa), but Kim *et al.* [KIM 12] observed values as high as $\theta \approx 80°$ for strongly salted (5 M NaCl) brine droplets left in the micromodel after drainage with CO_2 at 8.5 MPa and 45°C.

A possible reason for measurement discrepancies is contact angle hysteresis: any intermediate value between the advancing and receding angles can be observed by using the sessile drop method, unless specific procedures such as increasing or decreasing the pressure are adopted to shrink or swell the water drop. This difficulty can be circumvented by using the captive drop method, in which a drop of water (surrounded by the CO_2 phase) or a bubble of CO_2 (immersed in water) is held against the substrate and alternatively expanded and retracted by fluid injection and withdrawal to get the advancing and receding angles (see Figure 1.4). This method has been used by Chiquet *et al.* [CHI 07b] and Broseta *et al.* [BRO 12] to investigate the wetting behavior of a variety of substrates including quartz, muscovite mica, calcite, and a carbonate-rich rock (similar to that used in the breakthrough experiments presented in section 1.3 and Figures 1.1 and 1.2) over a large range of temperatures (up to 140°C), pressures (up to 150 bar), and brine salinities (up to NaCl saturation) representative of various geological storage conditions. In these measurements, a bubble of CO_2 was held beneath the substrate (immersed in water) and its volume varied by means of a capillary connected to a syringe (see Figure 1.4). The contact angle of interest, namely the drainage angle θ_{dr}, was observed to be low and not significantly altered under high CO_2 pressure [BRO 12]. While this needs to be further checked with other known water-wet minerals, we assume in the following that rock-forming minerals remain water-wet in the presence of CO_2 under geological storage conditions and incipient leakage (i.e. drainage) conditions (of CO_2 displacing water).

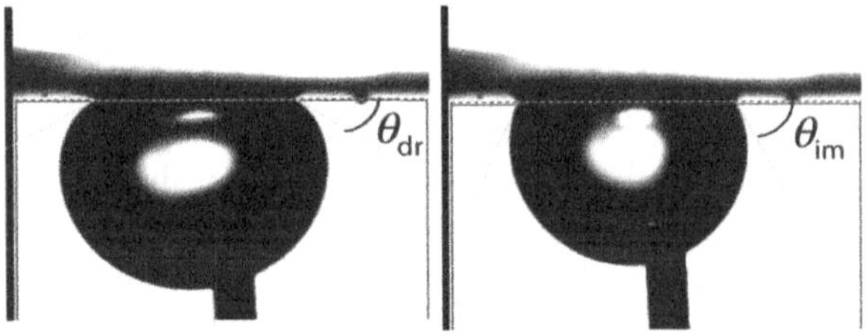

Figure 1.4. *Images of an expanding (left) and retracting (right) CO_2 bubble beneath muscovite mica in a low-salinity brine at 3.1 MPa and 35°C, giving respectively the drainage and imbibition angles θ_{dr} and θ_{im} [BRO 12]*

1.4. Maximum bottomhole pressure for storage in a depleted hydrocarbon reservoir

Acceptable bottomhole pressures are those that preserve the integrity of the seal rock: the maximum bottomhole pressure, which should not be exceeded during the CO_2 injection and storage process, corresponds to the breakthrough pressure of the seal rock. When the storage reservoir is a deep saline aquifer, the ability of the top low-permeability aquitard (or aquiclude) to confine a buoyant non-wetting fluid is not known in the first place, and the breakthrough pressure needs to be determined (see section 1.2). When the storage reservoir is a depleted hydrocarbon reservoir, the sealing capacity of the top seal is indeed proven up to the reservoir discovery pressure, but with respect to the hydrocarbons initially present in the reservoir (i.e. at reservoir discovery). It is not correct to assume that this seal will also be able to trap over the scheduled storage timescale (a few hundreds to thousands of years) of the CO_2-rich fluid that results from the injection process: refilling the depleted reservoir with CO_2 up to the reservoir discovery pressure will inevitably give rise to leakage through the top seal if the hydrocarbon reservoir was close to leakage conditions at discovery, because of the lower IFT between brine and the CO_2-rich fluid as compared to that between the brine and the hydrocarbon fluid [LI 06] – the drainage contact angles are supposed to be unchanged (see section 1.3).

A lower bound for the CO_2-rich fluid breakthrough (or displacement) pressure P_d (see equation [1.2]) is obtained from the following conservative hypothesis: the conditions for incipient hydrocarbon leakage by capillary failure were met at reservoir discovery, i.e. the discovery pressure P_{dis} is also the hydrocarbon breakthrough (or displacement) pressure, which thus exceeds the pressure P_w in the brine phase imbibing the seal rock by an amount proportional to $\gamma_{w,hc}\cos(\theta)/r$, where $\gamma_{w,hc}$ is the interfacial tension between brine and the hydrocarbon phase, r corresponds to the narrowest pore throats along the leakage pathway, and θ is the drainage contact angle. At some (small) distance from the reservoir boundary, the pore pressure P_w within the seal rock is the same as when the reservoir was discovered (because of the very low pressure diffusivity in low-permeable rocks).

Therefore, by using equation [1.2] and assuming that the drainage contact angles for the hydrocarbon- and CO_2-rich phases are equally low, the displacement pressure P_d is related to the discovery pressure P_{dis} as follows:

$$P_d - P_w < (P_{dis} - P_w)\ \gamma_{w,g}/\gamma_{w,hc} \qquad [1.4]$$

Since IFTs between water and hydrocarbons are at their maximum – equal to about 50 mN/m in geological storage conditions – for alkane-rich hydrocarbons, which is about twice the IFT between brine and CO_2 under similar conditions, the following simple rule of thumb can be enunciated: the pressure in a depleted hydrocarbon reservoir refilled with CO_2 should not exceed a value approximately half-way between the pore (i.e. brine) pressure in the seal rock and the reservoir discovery pressure. More precise estimates can be obtained from the IFTs between the formation brine and the hydrocarbons under the conditions of reservoir discovery and between the same brine and the CO_2-rich fluid present in the reservoir under storage conditions.

1.5. Evidences for capillary fracturing in seal rocks

"The view that gas breakthrough in water-saturated tight rocks leaves the porous skeleton intact is a simplification." The available experimental and theoretical evidences are that, in the course of the gas breakthrough process, some degree of rock deformation occurs that provides preferential pathways for gas migration; in addition, deformation often proceeds in a somewhat irreversible or quasi-permanent manner. The experimental evidences for the "formation, aperture and propagation of gas preferential pathways" [GAL 98] during gas breakthrough are numerous, starting with the early gas migration studies by Gallé, Harrington, and coworkers [GAL 98, HAR 99, HOR 99, GAL 00] in compacted clays (mostly bentonites) fully saturated with water. Intermittent gas outflow is apparent in some experiments, suggesting that "these pathways are relatively unstable" [GAL 98], and the gas breakthrough pressure needed to re-establish gas flow after a first gas breakthrough test (followed by flow arrest) is lower than the first gas breakthrough pressure (i.e. in the virgin clay), suggesting that the flow pathways opened under high gas pressure partially close when pressure falls [HOR 99]. Similar observations were made for gas (nitrogen) breakthrough experiments in the water-saturated Nordland shales overlying the Sleipner CO_2 storage site by Harrington *et al.* [HAR 09], who found some moderate alteration of the sealing capacity (i.e. a lower breakthrough pressure) upon repeat breakthrough measurements with the same sample.

In two recent studies of gas (CO_2 or N_2) breakthrough experiments [WOL 10, TON 10, TON 11] in various carbonate-rich rocks with permeabilities well below the micro-Darcy, a measurable loss in the rock sealing capacity (as quantified by a lower breakthrough over-pressure and a higher effective gas permeability) was observed again upon repetitive breakthrough experiments (in which rock samples were re-saturated with brine prior to starting a new breakthrough test).

In another recent study [SKU 12] of CO_2 breakthrough conducted in a micro-fractured shale (sampled from the Draupne formation in the North Sea) with permeability of a few nano-Darcy, the breakthrough pressure was observed to be sensitive to the confining and effective pressure conditions and to the compaction history, with a higher confining pressure and loading/unloading cycling of the sample leading to a higher breakthrough pressure; in addition, strain measurements indicated some pronounced dilation of the rock sample during breakthrough, presumably due to the decrease in effective pressure. Evidence for pressure-induced opening of micro-fractures by the intruding gas was provided by fitting the measured effective gas permeability versus volumetric strain to a simple model of fracture aperture [OLI 08].

In all the above-mentioned gas breakthrough experiments in clays or tight rocks, the alteration of the sealing capacity consecutive to gas breakthrough (i.e. the decrease in gas breakthrough pressure and the increase in clay or rock permeability) was moderate. However, these experiments are not numerous, and as mentioned by Harrington *et al.* [HAR 09], "a satisfactory explanation for how gas pathways develop (still) remains elusive." Clearly, some research effort should be devoted to the understanding of these phenomena, which have recently been referred to as "capillary fracturing" and they should be examined closely in non- or weakly-consolidated porous media, both numerically through 2D micro-mechanical simulations [JAI 09] and experimentally [HOL 12]. At small (pore) size, micro-fracturing or fracture opening arises from the difference in pressures between neighboring pores or grains when two different fluids (brine and gas) are present, and it is indeed more pronounced when capillary pressure forces are stronger, that is, for smaller grain sizes (i.e. smaller permeabilities) and larger brine/gas IFT. Capillary fracturing is ubiquitous in nature, and has been noted to occur in various contexts: gas venting from marine sediments, oil and gas generation and escape from the source rocks, or the vertical migration of gas through water-saturated tight sediments (e.g. mudstones) that is observed in many sedimentary basins [APL 99].

1.6. Summary and prospects

The typical procedure for assessing seal rock integrity with respect to an underlying over-pressurized gas consists of running gas breakthrough experiments, from which the gas breakthrough over-pressure and the effective gas permeability are determined. These two key parameters are related, respectively, to the maximum sustainable bottomhole pressure and to the gas leakage rate through the seal rock when this pressure is exceeded. These experiments are mandatory for the rocks (aquiclude or aquitard) overlying an aquifer used as a CO_2 storage reservoir, but not for the seal rock of a depleted hydrocarbon reservoir, where a lower bound for the maximum bottomhole pressure can be inferred from the discovery pressure of the reservoir and two IFTs: the IFT between brine and the hydrocarbons initially in place, and the IFT between brine and the stored CO_2-rich fluid.

The latter result holds provided the seal rock remains water-wet in the presence of CO_2, which seems to be the case if this rock was water-wet in the first place (i.e. in the presence of hydrocarbons). However, some seal rocks, such as mudstones, contain a non-negligible amount of organic matter [APL 11]: in this case, some significant wettability alteration in the presence of CO_2 is to be expected (see [BRO 12] for a review of the literature evidences) and the above simple approach might not hold. More generally, a critical issue is to understand and predict the distribution and connectivity of wettability and permeability heterogeneities within seal rocks [APL 11].

As emphasized in section 1.5, capillary fracturing is apparent in most gas breakthrough experiments, even though its (negative) effects on seal capacity have been shown to be rather limited in the cases investigated up to now. More work needs to be done in order to understand and predict the magnitude of these effects as a function of fluid properties and seal rock type and fracture mechanics. Another important topic that deserves to be investigated further with the above approaches and concepts is how fault seals (e.g. cataclastic fault rocks or permeability "baffles") function in the presence of CO_2 [TUE 12].

1.7. Bibliography

[ABR 93] ABRAMZON A.A., GAUKHBERG R.D., "Surface tension of salt solutions", *Russian Journal of Applied Chemistry*, vol. 66, pp. 1139–1146, 1315–1320, 1473–1479, 1643–1450, 1993.

[AGG 10] AGGELOPOULOS C.A., ROBIN M., PERFETTI E., VIZIKA O., "CO_2/$CaCl_2$ solution interfacial tensions under CO_2 geological storage conditions: influence of cation valence on interfacial tension", *Advances in Water Resources*, vol. 33, pp. 691–697, 2010.

[AGG 11] AGGELOPOULOS C.A., ROBIN M., PERFETTI E., VIZIKA O., "Interfacial tension between CO_2 and brine ($NaCl+CaCl_2$) at elevated pressures and temperatures: the additive effect of different salts", *Advances in Water Resources*, vol. 34, pp. 505–511, 2011.

[AMM 12] AMMAN-HILDENBRAND A., GHANISADEH A., KROOSS B.M., "Transport properties of unconventional gas systems", *Marine and Petroleum Geology*, vol. 31, pp. 90–99, 2012.

[APL 99] APLIN A.C., FLEET A.J., MACQUAKER J.H.S., *Introduction to "Muds and mudstones: physical or fluid flow properties"*, Geological Society, London, Special Publications 158, 1999.

[APL 11] APLIN A C., MACQUAKER J.H.S., "Mudstone diversity: origin and implications for source, seal and reservoir properties in petroleum systems", *AAPG Bulletin*, vol. 85, pp. 2031–2059, 2011.

[BIK 11a] BIKKINA P.K., SHOHAM O, UPPALURI R., "Equilibrated interfacial tension data of the CO_2-Water system at high pressures and moderate temperatures", *Journal of Chemical Engineering and Data,* vol. 56, pp. 3725–3733, 2011.

[BIK 11b] BIKKINA P.K., "Contact angle measurements of CO_2-water-quartz/calcite systems in the perspective of carbon sequestration", *International Journal of Greenhouse Gas Control*, vol. 5, pp. 1259–1271, 2011.

[BIK 12] BIKKINA P.K., "Contact angle measurements of CO_2-water-quartz/calcite systems in the perspective of carbon sequestration", *International Journal of Greenhouse Gas Control*, vol. 5, pp. 1259–1271, 2012.

[BOU 11] BOULIN P.F., BRETONNIER P., VASSIL V., SAMOUILLET A., FLEURY M., LOMBARD J.M., Entry pressure measurements using three unconventional experimental methods, Paper SCA 2011-02. http://www.scaweb.org/assets/papers/2011_papers/SCA2011-02.pdf

[BOU 12] BOULIN P.F., BRETONNIER P., GLAND N., LOMBARD J.M., "Contribution of the steady-state method to water permeability measurement in very low permeability porous media", *Oil and Gas Science and Technology*, vol. 67, pp. 387–401, 2012.

[BRO 12] BROSETA D. SHAH V., TONNET N., "Are rocks still water-wet in the presence of dense CO_2 or H_2S?", *Geofluids,* vol. 12, pp. 280–294.

[CAR 10] CARLES P., BACHAUD P., LASSEUR E., BERNE P., BRETONNIER P., "Confining properties of carbonated Dogger caprocks (Parisian Basin) for CO_2 storage purposes", *Oil and Gas Science and Technology,* vol. 65, pp. 461–472, 2010.

[CHA 09] CHALBAUD C., ROBIN M., LOMBARD J.M., MARTIN F., EGERMANN P., BERTIN H., "Interfacial tension and wettability evaluation for CO_2 geological storage", *Advances in Water Resources*, vol. 32, pp. 98–109, 2009.

[CHI 07a] CHIQUET P., DARIDON J.L., BROSETA D., THIBEAU S., "CO_2/water interfacial tensions under the pressure and conditions of geological storage", *Energy Conversion and Management*, vol. 48, pp. 736–744, 2007a.

[CHI 07b] CHIQUET P., BROSETA D., THUBEAU S., "Wettability alteration of caprock minerals by carbon dioxide", *Geofluids*, vol. 7, pp. 112–122, 2007b.

[DUC 12] DUCHATEAU C., BROSETA D., "A simple method for determining brine-gas interfacial tensions", *Advances in Water Resources*, vol. 42, pp. 30–36, 2012.

[ESP 10] ESPINOZA D.N., SANTAMARINA J.C., "Water-CO_2-mineral systems: interfacial tension, contact angle, and diffusion – Implications to CO_2 geological storage", *Water Resources Research*, vol. 46, pp. W07537, 2010.

[GAL 98] GALLÉ C., TANAI K., "Evaluation of gas transport properties of backfill materials for waste disposal: H_2 migration experiments in compacted Fo-Ca clay", *Clays and Clay Minerals*, vol. 46, pp. 498–508, 1998.

[GAL 00] GALLÉ C., "Gas breakthrough pressure in compacted Fo-Ca clay and interfacial gas overpressure in waste disposal context", *Applied Clay Science*, vol. 17, pp. 85–97, 2000.

[GEO 10] GEORGIADIS A., MAITLAND, G., TRUSLER J.P.M., BISMARCK A. "Interfacial tension measurements of the (H_2O+CO_2) system at elevated pressures and temperatures", *Journal of Chemical Engineering and Data*, vol. 55, pp. 4168–4175, 2010.

[GLU 94] GLUYAS J., SWARBICK, R., *Petroleum Geoscience*, Wiley-Blackwell, 1995.

[HAR 99] HARRINGTON J.F., HORSEMAN S.T., "Gas transport properties of clays and mudrocks", in APLIN A.C., FLEET A.J., MACQUAKER J.H.S. (eds), *Muds and Mudstones: Physical and Fluid Flow Properties*, Geological Society, London, Special Publications 158, pp. 107–124, 1999.

[HAR 09] HARRINGTON J.F., NOY D.J., HORSEMAN S.T., BIRCHALL D.J., CHADWICK R.A., "Laboratory study of gas and water flow in the Nordland Shale, Sleipner, North Sea", Chapter 30 in GROBE M., PASHIN J.C., DODGE R.J. (eds), *Carbon Dioxide Sequestration in Geological Media – State of the Science*, 1st. ed., AAPG Studies in Geology, vol 59, pp. 521–543.

[HIL 02] HILDENBRAND A., SCHLÖMER S., KROOSS B.M., "Gas breakthrough experiments on fine-grained sedimentary rocks", *Geofluids*, vol. 2, pp. 3–23, 2002.

[HIL 03] HILDENBRAND A., SCHLÖMER J.L., "Investigation of the morphology of pore space in mudstones – first results", *Marine and Petroleum Geology*, vol. 20, pp. 1185–1200, 2003.

[HIL 04] HILDENBRAND A., SCHLÖMER S., KROOSS B.M., LITTKE R., "Gas breakthrough experiments on pelitic rocks: comparative study with N_2, CO_2 and CH_4", *Geofluids*, vol. 4, pp. 61–80, 2004.

[HOL 12] HOLTZMANN R., SZULCZEWSKI M.L., JUANES R., "Capillary fracturing in granular media", *Physical Review Letter*, vol. 108, pp. 264504, 2012.

[HOR 99] HORSEMAN S.T., HARRINGTON J.F., SELLIN P., "Gas migration in clay barriers", *Engineering Geology*, vol. 54, pp. 139–149, 1999.

[JAI 09] JAIN A.K., JUANES R., "Preferential mode of gas invasion in sediments: mechanistic model of coupled multiphase fluid flow and sediment mechanics", *Journal of Geophysical Research*, vol. 114, pp. B08101, 2009.

[JUN 12] JUNG J., WAN J., "Supercritical CO_2 and ionic strength effects on wettability of silica surfaces: contact angle measurements", *Energy and Fuels*, vol. 26, pp. 6053–6059, 2012.

[KIM 12] KIM Y., WAN J., KNEAFSEY T.J., TOKUNAGA T.K., "Dewetting of silica surfaces upon reactions with supercritical CO_2 and brine: pore-scale studies in micromodels", *Environmental Technology*, vol. 46, pp. 4228–4235, 2012.

[LI 05] LI S., DONG M., LI Z., HUANG S., QING H., NICKEL E., "Gas breakthrough pressure for hydrocarbon seal rocks: implications for the security of long-term CO_2 storage in the Weyburn field", *Geofluids*, vol. 5, pp. 326–334, 2005.

[LI 06] LI Z., DONG M., LI Z., HUANG S., "CO_2 sequestration in depleted oil and gas reservoirs-caprock characterization and storage capacity", *Energy Conversion and Management*, vol. 47, pp. 1372–1382, 2006.

[LI 12] LI X., BOEK E., MAITLAND G.C., TRUSLER J.P.M., "Interfacial tension of (brines+CO_2): 0.864 NaCl + 0.136 KCl at temperatures between (298 and 448) K, pressures between (2 and 50) MPa, and total molalities of (1 to 5) mol.kg^{-1}", *Journal of Chemical Engineering and Data*, vol. 57, pp. 1369–1375, 1078–1088, 2012.

[MIL 12] MILLS J., RIAZI J., SOHRABI M., *Wettability on Common Rock-Forming Minerals in a CO_2-Brine System at Reservoir Conditions*, SCA, Austin TX, 2012.

[NEW 04] NEWSHAM K.E., RUSHING J.A., LASSWELL P.M., COX J.C., BLASINGAME T.A., "A comparative study of laboratory techniques for measuring capillary pressures in tight sands", Paper SPE 89866, *SPE Annual Technical Conference and Exhibition*, Houston, TX, pp. 26–29, September 2004.

[OLI 08] OLIVELLA S., ALONSO E.E., "Gas flow through clay barriers", *Geotechnique*, vol. 58, pp. 157–176, 2008.

[REN 00] REN Q.Y., CHEN G.J., YAN W., GUO T.M., "Interfacial tension of (CO_2 + CH_4) + water from 298 K to 373 K and pressures up to 30 MPa", *Journal of Chemical Engineering Data*, vol. 45, pp. 610–612, 2000.

[RUT 12] RUTQVIST J., "The geomechanics of CO_2 storage in deep sedimentary formations", *Geotechnology Geological Engineering*, vol. 30, pp. 525–551, 2012.

[SHA 08] SHAH V., BROSETA D., MOURONVAL G., MONTEL F., "Water/acid gas interfacial tensions and their impact on acid gas geological storage", *International Journal of Greenhouse Gas Control*, vol. 2, no. 4, pp. 94–604, 2008.

[SKU 12] SKURTVEIT E., AKER E., SOLDAL M., ANGELI M., WANG Z., "Experimental investigation of CO_2 breakthrough and flow mechanisms in shale", *Petroleum Geoscience*, vol. 18, pp. 3–15, 2012.

[THO 68] THOMAS L.K., KATZ D.L., TEK M.R., "Threshold pressure phenomena in porous media", *SPE Journal*, vol. 8, pp. 174–184, 1968.

[TON 10] TONNET N., BROSETA D., MOURONVAL G., "Evaluation of the petrophysical properties of a carbonate-rich caprock for CO_2 geological storage purposes", *SPE 131525, SPE EUROPEC/EAGE Annual Conference and Exhibition*, Barcelona, Spain, pp. 14–17, June 2010.

[TON 11] TONNET N., MOURONVAL G., CHIQUET P., BROSETA D., "Petrophysical assessment of a carbonate-rich caprock for CO_2 geological storage purposes", *Energy Procedia*, vol. 4, pp. 5422–5429, 2011.

[TUE 12] TUECKMANTEL C., FISHER Q.J., MANZOCCHI, SKACHKOV S., Grattoni C.A., "Two-phase fluid flow properties of cataclastic fault rocks; implications for CO_2 storage in saline aquifers", *Geology,* vol. 40, pp. 39–42, 2012.

[WAN 12] WANG S., EDWARDS I.M., CLARENS A.F., "Wettability Phenomena at the CO_2–Brine–Mineral Interface: Implications for Geologic Carbon Sequestration", *Environmental Science and Technology*, 2012 (in press).

[WAT 87] WATTS N.L., "Theoretical aspects of caprock and fault seals for single- and two-phase hydrocarbon columns", *Marine and Petroleum Geology*, vol. 4, pp. 274–307, 1987.

[WOL 10] WOLLENWEBER J., ALLES S., BUSCH A., KROOSS B.M., STANJEK H., LITTKE R., "Experimental investigation of the CO_2 sealing efficiency of caprocks", *International Journal of Greenhouse Gas Control*, vol. 4, pp. 211–241, 2010.

[YAN 01] YAN W., ZHAO G.Y., CHEN G.J., GUO T.M., "Interfacial tension of (methane+nitrogen)+water and (carbon dioxide +nitrogen)+water Systems", *Journal of Chemical Engineering and Data,* vol. 46, pp. 1544–1548, 2001.

[YAN 08] YANG D., GU Y., TONTIWACHWUTHIKUL P., "Wettability determination of the reservoir brine-reservoir rock system with dissolution of CO_2 at high pressures and elevated temperatures", *Energy and Fuels*, vol. 22, pp. 504–509, 2008.

[ZWE 04] ZWEIGEL P., LINDENBERG E., MOEN A., WESSEL BERG D., "Towards a methodology for top seal efficacy assessment for underground CO_2 storage", *7th International Conference on Greenhouse Gas Control Technologies (GHGT-7)*, Vancouver, Canada, 2004.

Chapter 2

Gas Migration through Clay Barriers in the Context of Radioactive Waste Disposal: Numerical Modeling of an *In Situ* Gas Injection Test

2.1. Introduction

In carbon dioxide (CO_2) storage problems as well as in the field of radioactive waste confinement, the question of gas transfers in argillaceous formations is a crucial issue. The geological sequestration of CO_2 may induce gas transport in the caprock of a depleted gas reservoir, and we must ensure that such CO_2 geological disposal activities are safe and do not induce any undesired environmental impact. However, during long-term repository of radioactive waste in clayey rocks, a large amount of gases will be produced by the deterioration of the disposal components (packages, packaging components, metallic construction material). This production of gas could affect the safety function of the clay barriers. The understanding of their production and migration mechanisms in the host rock is of particular significance for the safety assessment of such solutions.

As a result, numerous laboratory experiments for gas injection have been performed on clayey materials [VOL 95, ORT 97, GAL 00, HOR 99, HAR 03, HIL 02, ARN 08] exhibiting complex behavior, in particular, for gas flow under water saturated conditions. Long-term gas injection tests on argillaceous materials show flow

---

Chapter written by Pierre GÉRARD, Jean-Pol RADU, Jean TALANDIER, Rémi de La VAISSIÈRE, Robert CHARLIER and Frédéric COLLIN.

instabilities, which are interpreted as an indication of the opening and the closing of localized gas pathways through the samples [HOR 99, HAR 03, MAR 05].

However, large-scale tests have rarely been conducted [MAR 05, CUS 11]. Examination of these different data, nevertheless, also shows evidence of the development of conductive pathways, as at the laboratory scale. A new large-scale gas injection test has recently been performed at the Meuse/Haute-Marne Underground Research Laboratory (URL) in France. This well-monitored experiment, called PGZ1, studies the migration of nitrogen in the host rock, i.e. Callovo–Oxfordian claystone [VAI 12]. Different sensors located in the host rock provide the pore pressure evolution with the gas injection conditions. This chapter is devoted to the modeling of this experiment and the analysis of the gas migration mechanisms.

The modeling of experiments studying gas migration in clay materials is a complex task. The experimental observations do not generally conform to standard concepts of two-phase flow. Different approaches have thus been proposed to consider the gas flow instabilities. For instance, the lattice models view the porous medium as the superimposition of capillary networks with different radius and density [GRI 94, XU 97a, XU 97b]. In finite element code, the development of preferential pathways could be reproduced from the initial and natural heterogeneity of the material [DEL 02]. A heterogeneous distribution of the porosity or the permeability allows the reproduction of some experimental data obtained on a gas injection test. A final way to reproduce the generation of discrete paths is the introduction of additional hydromechanical coupling between the permeability, the water retention curve, and the strains [OLI 08, GER 11, GER 12]. The aperture of discrete paths is then the main variable to account for permeability and capillary pressure variations. Even though all these methods can be used to reproduce and explain the experimental observations, they do not constitute a predictive model, which is needed in safety assessment for long-term radioactive waste repository.

In this chapter, we propose to deal with the modeling of the PGZ1 *in situ* experiment with two-phase flow model, in order to highlight how a predictive approach could reproduce the experimental observations coming from a gas injection field test. The geometry of the problem and the permeability anisotropy lead us to perform 3D modeling, which requires a tremendous computational effort. However, the hydromechanical coupling is usually limited in such modeling of gas problems if two-phase flow and perfect plasticity are assumed [GER 08]. Its influence could be neglected and hydraulic modeling of the experiment is thus achieved to reduce the degrees of freedom of the problem. The purpose of this contribution focuses on the role of each component of the system on the gas transfer mechanisms. It highlights the most significant characteristics to take into account in a numerical approach of a gas injection problem.

In this chapter, the PGZ1 experiment is first described and the experimental observations are detailed. The boundary value problem is then presented and the two-phase flow model is recalled. The numerical results show the ability of the two-phase model to reproduce large-scale experiments.

2.2. Field experiment description

The PGZ1 *in situ* experiment is devoted to the characterization of the gas transfer properties of Callovo–Oxfordian claystone, thanks to a gas injection test performed from an experimental borehole. This experiment takes place in the Meuse/Haute-Marne Underground Research Laboratory (URL) in France between the GED and GEX horizontal galleries that are parallel experimental galleries spaced apart by 25 m.

From the GED gallery, two 28-m long parallel and downward boreholes are drilled with a dip of 35° to the horizontal plane and a strike perpendicular to the GED gallery axis (Figure 2.1). These two boreholes have a diameter of 76 mm and are composed of three intervals with pore pressure sensors separated by packers. The gas injection in host rock is performed from the central interval of the injection borehole. The other intervals are used to follow the pore pressure evolution in the rock mass.

The last 23 m-long downward borehole is drilled from the end of the GEX gallery with the same strike as that of the GEX gallery. The dip of this borehole is 47° to the horizontal plane. Its diameter is 101.3 mm. It is used for the monitoring of the displacements in argillite, thanks to extensometers. The shortest distance between the extensometer borehole and the center of the gas injection interval is equal to 1.132 m.

After the setup of the sensors and packers, the pore pressure evolution is monitored over 188 days, before the beginning of the nitrogen injection. The gas test comprises different periods of controlled gas flow injection rate, interrupted by "shut-in" phases when gas injection ceases (Figure 2.2).

In this contribution, we will analyze the measured pore pressure evolution, thanks to six sensors located in the different intervals of the injection and measuring boreholes (Figure 2.3). After the drilling of the borehole and the setup of the sensors, a re-equilibrium phase is observed. The pore pressure increases and reaches the initial water pressure in the rock mass around 4.5 MPa (Figure 2.3). Two hydraulic tests (pulse tests) are then performed in the interval 1 of the injection borehole in order to characterize the permeability of the host rock. At the same time, a drop in the pore pressures is observed for all the intervals of the two parallel boreholes. A strong correlation between the distance to the third borehole and the kinetics and amplitude of the pressure reduction in the different intervals leads us to interpret this decrease in interstitial pressure as an effect of drainage induced by the third borehole. This borehole is indeed filled with a cemented material that presents a hydraulic conductivity (10^{-16} m^2), which is three orders of magnitude higher than the sound argillite. This explanation has been validated by 3D modeling of the re-equilibrium phase, but is not presented in this contribution.

When the gas test begins with controlled nitrogen flow rate periods, interrupted by "shut-in" phases, an increase of the pore pressure is observed in the interval 1 of the injection borehole. The effect of the gas injection is not detected in the other intervals, except for the central interval of the measuring borehole (interval 4) but with smaller magnitude (Figure 2.3(c)).

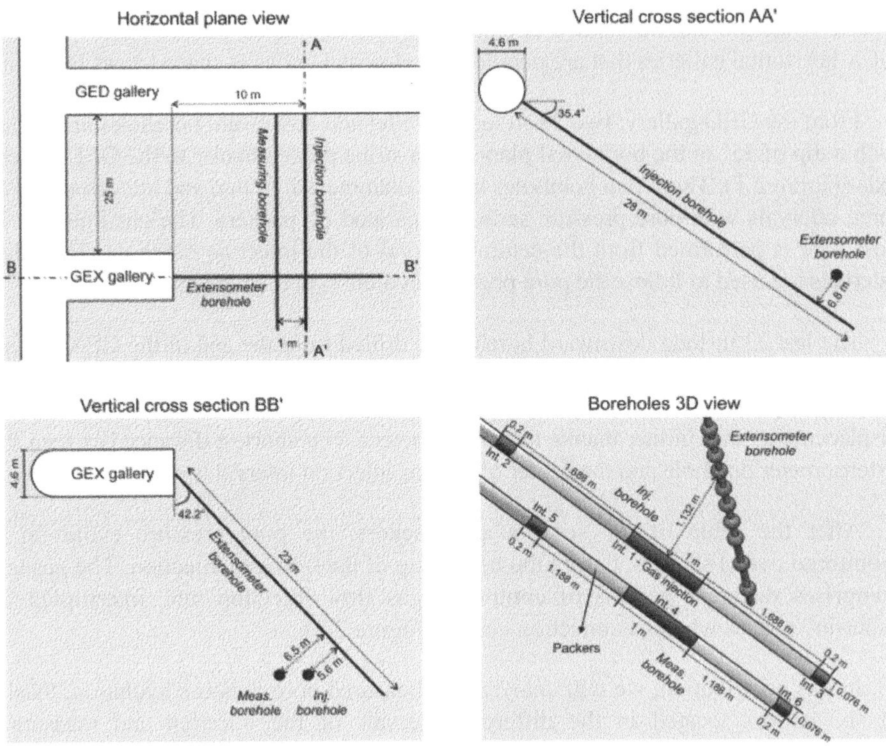

Figure 2.1. *Schematic position of the pore pressure sensors (distances are measured from the gallery wall)*

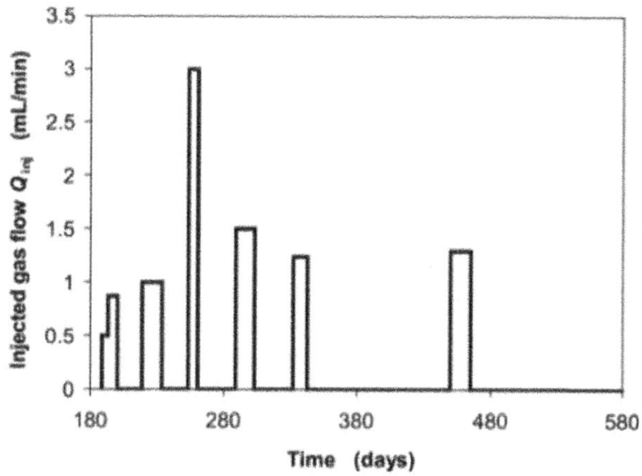

Figure 2.2. *Gas flow injected experimentally at one face of the injection interval*

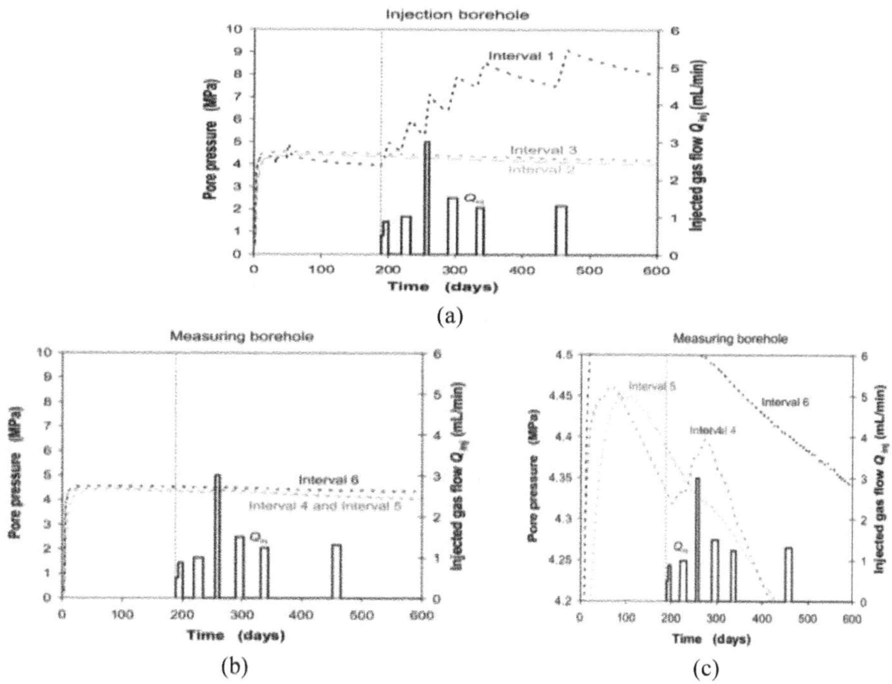

Figure 2.3. *Time evolution of pore pressures in sensors before and during gas injection tests. a) Injection borehole; b) and c) measuring borehole without and with zoom*

2.3. Boundary value problem

A hydraulic modeling of the problem is performed with the finite element code Lagamine. The drilling of the injection borehole followed by a period where the pore pressures evolved freely at the borehole wall is first modeled in 188 days. The gas problem is not solved during these two steps. The nitrogen injection is then modeled through a controlled gas flow corresponding to the experimental model (Figure 2.2).

In this chapter, we focus on the modeling of the gas injection phases. In order to clearly highlight the influence of each component of the system on the response of the test, 1D modeling is first proposed. In the second step, 3D modeling is performed and provides additional information on the behavior of the rock mass during the gas injection. 3D modeling is indeed needed to obtain convincing numerical results, because the injection borehole is an inclined borehole and argillite presents anisotropy of the permeability.

2.3.1. *1D and 3D geometry and boundary conditions*

1D axisymmetric modeling of the PGZ1 experiment is first achieved. An injection interval with a radius of 3.8 cm is considered, followed by the rock mass. In the second step, a half 3D mesh is obtained by rotation around the axis of the injection borehole. The injection interval (interval 1) is also modeled (Figure 2.4). The volume of the injection interval will be discussed in section 2.4 which is devoted to the numerical results, because some uncertainties remain on this question.

In 1D and 3D problems, the outer boundaries are located far enough in order to impose the initial water and gas pressures there. The initial conditions in the injection interval will be discussed in the section devoted to the numerical results. The initial water pressure in the rock mass is defined at the beginning of the borehole drilling and is equal to the hydrostatic pressure (4 MPa). At the end of the resaturation phase, the water pressure returns to this initial value about 4 MPa and the rock mass is saturated. The gas injection is modeled by imposing the experimental gas flow at one lateral face of the injection interval.

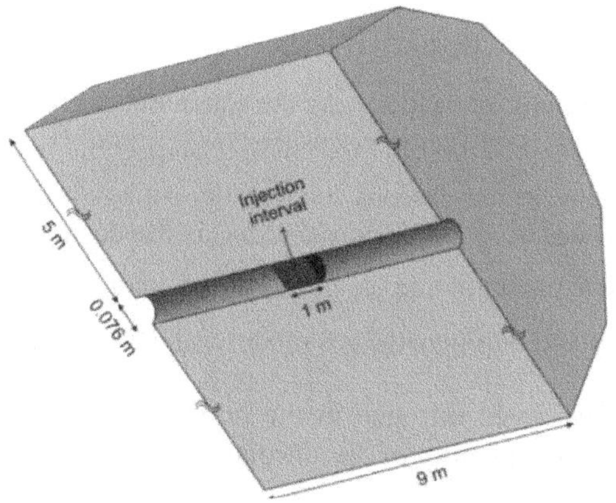

Figure 2.4. *Schematic view and geometry of the 3D mesh and the outer boundaries*

2.3.2. *Hydraulic model*

To reproduce water and nitrogen transfers in partially saturated porous media, a two-phase flow model is used. This model comprises a liquid phase, consisting of liquid water and dissolved nitrogen and a gaseous phase, which is an ideal mixture of dry nitrogen and water vapor. It takes into account the advection of each phase using the Darcy's law ($\underline{q}_l$ and $\underline{q}_g$) and the diffusion of the components within each phase using the Fick's law ($\underline{i}_v$, $\underline{i}_{N_2}$ and $\underline{i}_{N_2-d}$). All the details of the two-phase flow models are available in [CHA 12], but the different flows are defined hereafter:

$$\underline{q}_l = -\frac{\underline{\underline{K}}_w^{sat}\, k_{r,w}}{\mu_w}\underline{grad}(p_w) \qquad [2.1]$$

$$\underline{q}_g = -\frac{\underline{\underline{K}}_g^{dry}\, k_{r,g}}{\mu_g}\underline{grad}(p_g) \qquad [2.2]$$

$$\underline{i}_v = -\phi(1-S_{r,w})\tau D_{v/N_2}\rho_g\,\underline{grad}\left(\frac{\rho_v}{\rho_g}\right) = -\underline{i}_{N_2} \qquad [2.3]$$

$$\underline{i}_{N_2-d} = -\phi S_{r,w}\tau D_{N_2-d/w}\rho_w\,\underline{grad}\left(\frac{\rho_{N_2-d}}{\rho_w}\right) \qquad [2.4]$$

where $\underline{\underline{K}}_w^{sat}$ and $\underline{\underline{K}}_g^{dry}$ are, respectively, the water and gas permeabilities tensor in saturated and dry conditions; $k_{r,w}$ and $k_{r,g}$ are the water and gas relative permeabilities; μ_w and μ_g are the water and gas dynamic viscosities; D_{v/N_2} and $D_{N_2-d/w}$ are the diffusion coefficients respectively in the gaseous mixture dry nitrogen–water vapor and for the dissolved nitrogen in water; ϕ is the porosity; $S_{r,w}$ is the degree of saturation; τ is the tortuosity of the porous medium; and ρ_v, ρ_{N_2-d}, ρ_g and ρ_w are the volumetric mass of the vapor, the dissolved nitrogen, the gas mixture, and the liquid water, respectively.

The amount of dissolved nitrogen in the liquid phase is always in equilibrium and is proportional to the quantity of dry nitrogen (Henry's law):

$$\rho_{N_2-d} = H_{N_2}(T)\rho_{N_2} \quad [2.5]$$

where H_{N_2} is Henry's coefficient for dissolved nitrogen, depending on temperature.

2.3.3. *Hydraulic parameters*

To reproduce the unsaturated behavior of the rock mass, a retention curve and the water and gas permeability curves have to be defined. A review of experimental data available on unsaturated characteristics of Callovo–Oxfordian claystone is presented in [CHA 12]. Based on that, a van Genuchten's retention curve is used [VAN 80], with parameters calibrated in order to reproduce the experimental results at best. Similarly, a van Genuchten water relative permeability relationship is defined, while a cubic law is used for gas relative permeability:

$$S_{r,w} = \left(1 + \left(\frac{p_c}{P_r}\right)^n\right)^{\frac{1}{n}-1} \quad [2.6]$$

$$K_w = K_w^{sat} k_{r,w} = K_w^{sat} \sqrt{S_{r,w}} \left(1 - \left(1 - S_{r,w}^{1/m}\right)^m\right)^2 \quad [2.7]$$

$$K_g = K_g^{dry} k_{r,g} = K_g^{dry} \left(1 - S_{r,w}\right)^3 \quad [2.8]$$

with p_c being the capillary pressure, P_r and n the two parameters of the retention curve, and m is a parameter indicating the water and gas relative permeability relationship (usually considered as $m = 1 - 1/n$).

The hydraulic parameters of argillite and the injection interval introduced in the two-phase flow model are defined in Table 2.1. A permeability anisotropy ratio equal to 3 is considered, which corresponds to previous observations from [AND 05].

The injection interval is modeled by considering an equivalent porous media, with high permeability, porosity equal to 1, and a low air entry pressure. No water and gas relative permeability curve is defined for this component.

		Undisturbed argillite	
		1D	3D
$K_{w,hor}^{sat}$	Horizontal water permeability (m²)	4×10^{-20}	4×10^{-20}
$K_{w,vert}^{sat}$	Vertical water permeability (m²)		1.33×10^{-20}
$K_{g,hor}^{dry}$	Horizontal gas permeability (m²)	4×10^{-18}	4×10^{-18}
$K_{g,vert}^{dry}$	Vertical gas permeability (m²)		1.33×10^{-18}
ϕ	Porosity (-)	0.18	
τ	Tortuosity (-)	0.25	
P_r	van Genuchten air entry pressure (MPa)	15	
n	van Genuchten parameter (-)	1.49	
m	van Genuchten parameter (-)	0.55	

Table 2.1. *Hydraulic parameters*

2.4. Numerical results

The numerical results of the drilling phase followed by the increase of the pore pressures in the boreholes are not the purpose of this chapter. Nevertheless, a 3D modeling of these phases (not presented here) has shown that the drainage induced by the extensometer borehole could explain the decreasing trend of the pore pressures observed experimentally in the sensors located along the injection and measuring boreholes (Figure 2.3) and are discussed in section 2.2.

Therefore, this section is devoted to the hydraulic modeling of the gas injection test. First, we focus on the numerical results obtained in the injection interval (interval 1), where the increase of the pore pressures is the most significant. As the drainage of the extensometer borehole is not taken into account in the modeling of the gas injection phase, a correction of the experimental data is performed to cancel the decreasing trend observed on the pore pressure evolutions.

2.4.1. *1D modeling*

1D modeling is first performed with the geometry and the boundary conditions defined in the above mentioned sections.

From an experimental point of view, some uncertainties remained on the evaluation of the exact volume of the injection interval. The length and the radius of the interval are well known (L = 1 m and R = 38 mm), but the injection interval comprises a steel internal tube, surrounded by an annular chamber. Only this last component takes part in the gas transfer into the rock mass. Moreover, the volume of the feed lines has to be determined accurately, because it contributes to the injection volume. Furthermore, the presence of breakouts at the borehole wall and possible convergence of the borehole between drilling and the gas injection could modify the interval volume.

The numerical results presented hereafter will show the importance of the injection volume determination. Two injection volumes have indeed been considered. First, a volume of the injection interval of 4.536 L is defined, which corresponds to the volume of a theoretical cylindrical interval, with a length of 1 m and a diameter equal to 76 mm (Figure 2.5). In the second step, a volume of 1.040 L is considered. It is deduced experimentally from the previous water injection tests and allows the determination of the total injection volume (annular interval volume, feed lines volume, and the breakouts along the borehole before the gas injection steps).

In the first modeling, a volume of 4.536 L is thus considered. The injection interval is assumed initially saturated at the beginning of the gas injection, with a water pressure equal to the initial value in the rock mass (4 MPa). The water and gas pressures obtained in the injection interval are compared with the experimental measurements of the pore pressure in Figure 2.5(a). Water and gas pressures evolve similarly, which infers that initial water located in the interval is not totally pushed into the rock mass at the end of the gas injection test. It is confirmed by the analysis of the degree of saturation that a small amount of water remains in the annular chamber (Figure 2.5(c)).

On the other hand, a more accurate estimation of the injection volume (1.040 L) is used. The results of this second modeling show that the injection interval is totally desaturated from the third injection peak (about 280 days), because a smaller water volume was initially available in the injection interval (Figures 2.5(b)–(d)). When water is totally pushed from the interval to the rock mass, gas and water pressures evolve separately, which leads to an increase of the gas pressure. It improves the comparison with the experimental data for long-term predictions. In this second case, a low desaturation of argillite is observed near the borehole wall due to the nitrogen transfers (Figure 2.5(d)). These first results highlight how a correct estimation of the injection volume is needed and influence the numerical results.

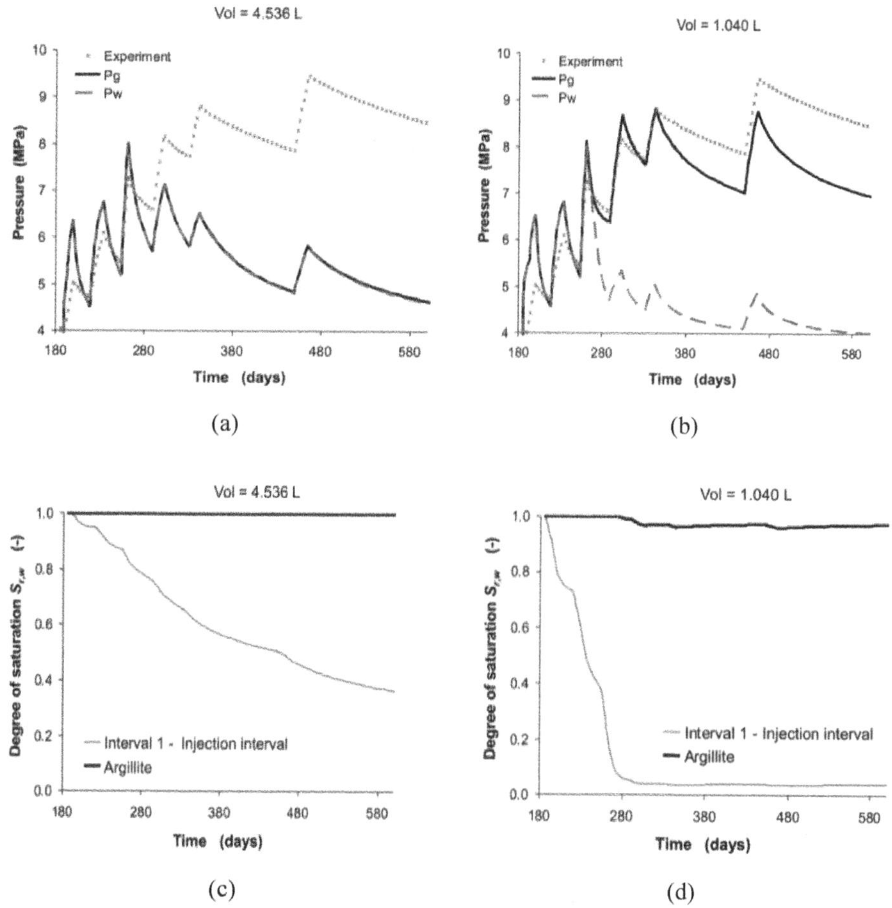

Figure 2.5. *Time evolution of pore pressures in the injection interval. Comparisons between experimental and numerical results for an injection interval volume of a) 4.536 L and b) 1.040 L. Time evolution of degree of saturation in the injection interval and in argillite at the wall for an injection interval volume of (c) 4.536 L and (d) 1.040 L*

Moreover, the assumption of an initially saturated interval has to be discussed. A volume of the residual water was indeed removed from the injection interval just before the nitrogen injection [VAI 12]. On the basis of the volume extracted, the initial degree of saturation in the interval can be deduced. An initial degree of saturation of 0.221 is thus assumed in the interval for the modeling. It induces a compressibility increase of the nitrogen–vapor mixture and lower pore pressures are thus obtained during the first injection peaks (Figure 2.6). It is also easier to inject nitrogen and to remove water from the interval if the initial water volumes considered are lower. In particular the numerical results are strongly improved during the first two peaks when the correct initial degree of saturation is considered. For long-term predictions, the initial conditions in the interval do not influence the results, because all the water has already been removed from the interval, whatever the initial saturation. Therefore, the measurements of pore pressures are only influenced by gas transfers in the rock mass during the last steps.

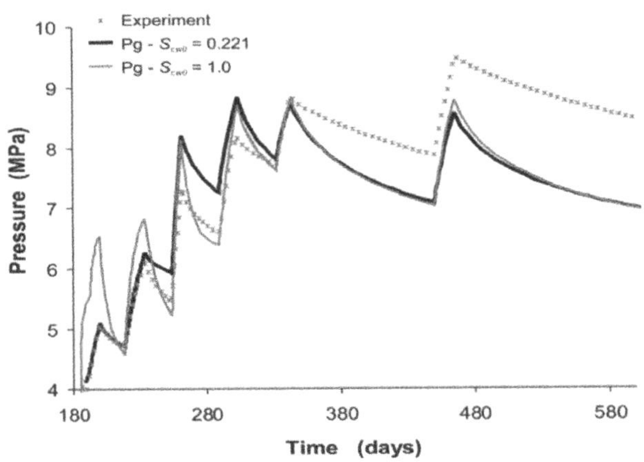

Figure 2.6. *Influence of the initial degree of saturation in the injection interval on the time evolution of pore pressures in the injection interval – comparisons between experimental and numerical results*

In order to improve the comparison with the pore pressure measurements, the rock mass behavior now has to be characterized more precisely. During the excavation process of the borehole, a perturbed zone is usually created around the opening [BOS 02, TSA 05, BLU 07, LEV 10]. This excavation damaged zone (EDZ) is a zone in which hydromechanical modifications induce major changes in flow and transport properties, resulting in macro- and micro-fracturing and a rearrangement of rock structures. It could lead to a significant increase of the permeability and also to a modification of the retention properties of the host rock, which probably influences the results of the gas injection test.

The objective of this chapter is not to propose a numerical approach to tackle the development of fractures during the drilling and its consequence in terms of transfer properties, but to focus on the influence of each component of the system on the gas flows. For this reason, the presence of an excavated damaged zone is initially assumed in our numerical problem. The extent of the EDZ is assumed close to the radius of the borehole (4 cm), which corresponds to previous experimental observations [BOS 02, BLU 07, SHA 08]. In this disturbed zone, a constant permeability is considered 500 times higher than the undisturbed claystone permeability. The same ratio is applied to the water and gas permeabilities ($K_{w,EDZ}^{sat} = 2 \cdot 10^{-17} \text{m}^2$ and $K_{g,EDZ}^{dry} = 2 \cdot 10^{-15} \text{m}^2$). The retention properties are also modified by the excavation, and the parameter P_r of the van Genuchten's retention curve is decreased due to the micro-fracturing of the rock mass ($P_{r,EDZ} = 3$ MPa).

The introduction of a damaged zone at the borehole wall improves strongly the agreement with the experimental observations, as shown in Figure 2.7. The experimental data are now well reproduced during the first four injection peaks. Moreover, the modification of the damaged zone permeability does not influence the predictions, which demonstrates that the gas transfers around the borehole are especially controlled by the retention characteristics of the porous media and not by its permeability. The modification of the retention characteristics induces indeed the development of a desaturated zone. A gaseous phase is thus created in the porous medium, which makes the nitrogen migration easier.

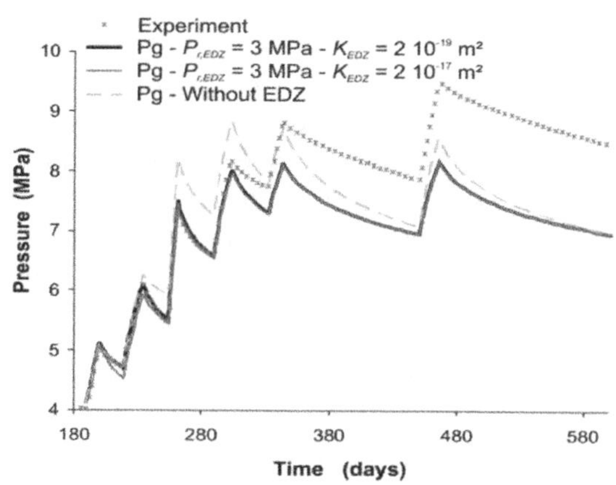

Figure 2.7. *Influence of the introduction of an excavated damage zone on the time evolution of pore pressures in the injection interval – comparisons between experimental and numerical results*

The consideration of the correct initial conditions of the injection interval and the introduction of an excavated damaged zone has allowed a good reproduction of the experimental data during the first four injection steps. Nevertheless, long-term predictions should be still improved. We could consider that the last injection peaks test particularly the sound claystone characteristics, because it has to be shown that the pressures at the end of the experiment are not really influenced by the characteristics of the damaged zone.

The hydraulic characteristics of sound Callovo–Oxfordian argillite have already been determined in numerous laboratory or field experimental studies [AND 05, DIS 07, PHA 07, HOM 04, ESC 05, HEI 02]. Water and gas permeability in saturated and dried conditions, respectively, are usually well known as the parameters of the retention curve based on the van Genuchten's relationship. On the other hand, some uncertainties remain in the evolution of the gas permeability with the degree of saturation. Although some experimental studies have been performed on this issue [BOU 08a, BOU 08b, YAN 08, ZHA 04], few of them investigate the gas transfer properties of argillite close to the saturation in view of the experimental difficulties to impose accurately high degree of saturation. Cubic law is thus usually considered as gas relative permeability function even if we could not verify its relevance in the quasi-saturated domain, but other relationships as that of the Parker's exist:

$$K_g = K_g^{dry} k_{r,g} = K_g^{dry} \sqrt{1-S_{r,w}} \left(1-S_{r,w}^{1/\lambda}\right)^{2\lambda} \qquad [2.9]$$

where λ is a parameter.

If the Parker's relationship (with $\lambda = 1.6$) is used to define the gas relative permeability not only in the sound rock but also in the excavated damaged zone, a better reproduction of the experimental pore pressures is obtained (Figure 2.8(a)). The Parker's function provides lower gas permeability than the cubic law (Figure 2.8(b)), which allows the increase of the gas pressure during the last two injection steps. The analysis of the saturation profiles shows not only the desaturation of the damaged zone, but also of the undisturbed host rock during the last peaks.

From this last satisfactory 1D modeling, we could extract a set of parameters characterizing the behavior of the injection interval, the excavated damaged zone, and the undisturbed rock. These parameters will be used in the 3D modeling of the problem.

2.4.2. *3D modeling*

The aim of the 3D modeling is to highlight the role of the permeability anisotropy and the extent of the damaged zone along the borehole and the role of the axial flows. In 1D modeling, the disturbed zone is assumed just in front of the injection interval. It is certainly not relevant to extend this disturbed zone all around the injection borehole, because the three intervals are separated by packers (Figure 2.1) for which the swelling pressures are higher than the gas pressures. Two domains are therefore defined in the EDZ along the injection borehole in front of the intervals or in front of the packers. The same radial extents as that in the 1D modeling are assumed for the EDZ around the borehole (4 cm), but its characteristics differ according to the domains. In front of the three intervals, the parameters coming from the 1D modeling are introduced in the two-phase flow model. On the other hand, we assume that the swelling of the packers allows recovering the initial characteristics of the sound rock in front of the packers. A summary of the hydraulic parameters introduced in the 3D modeling is presented in Table 2.2.

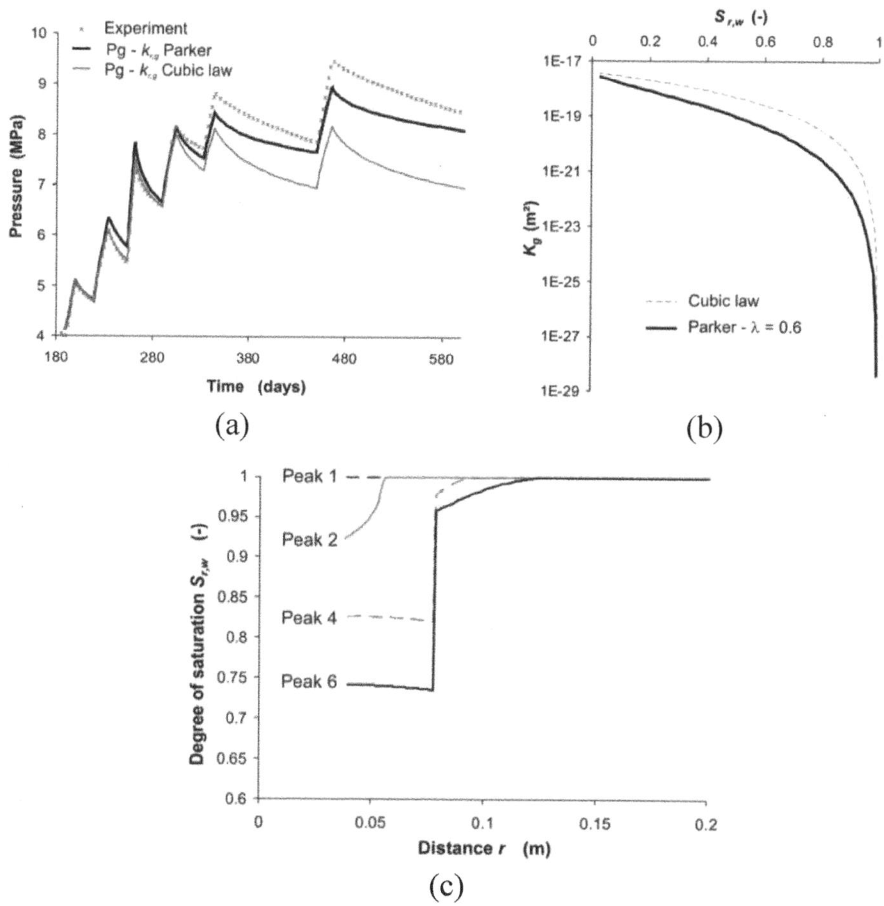

Figure 2.8. *a) Influence of the gas relative permeability of the undisturbed argillite on the time evolution of pore pressures in the injection interval – comparison between experimental and numerical results. b) Gas permeability against degree of saturation. c) Degree of saturation profiles at different gas injection peaks*

		Undisturbed argillite	EDZ – Interval	EDZ – Packer
$K_{w,hor}^{sat}$	Horizontal water permeability (m²)	$4{\cdot}10^{-20}$	$2{\cdot}10^{-19}$	$4{\cdot}10^{-20}$
$K_{w,vert}^{sat}$	Vertical water permeability (m²)	$1.33{\cdot}10^{-20}$	$6.67{\cdot}10^{-20}$	$1.33{\cdot}10^{-20}$
$K_{g,hor}^{dry}$	Horizontal gas permeability (m²)	$4{\cdot}10^{-18}$	$2{\cdot}10^{-18}$	$4{\cdot}10^{-18}$
$K_{g,vert}^{dry}$	Vertical gas permeability (m²)	$1.33{\cdot}10^{-18}$	$6.67{\cdot}10^{-18}$	$1.33{\cdot}10^{-18}$
ϕ	Porosity (-)	0.18	0.18	0.18
τ	Tortuosity (-)	0.25	0.25	0.25
P_r	van Genuchten air entry pressure (MPa)	15	3	15
n	van Genuchten parameter (-)	1.49	1.49	1.49
m	van Genuchten parameter (-)	0.55	0.55	0.55
λ	Parker parameter (-)	1.6	1.6	1.6

Table 2.2. *Hydraulic parameters for the 3D modeling*

The 3D modeling allows the comparison between the experimental and numerical results in the six intervals where measurements of pore pressures are performed. In the injection interval, the results are similar to those obtained in the 1D modeling, as shown in Figure 2.9. A good agreement with the experimental observations is thus obtained. The desaturation of the injection interval seems to be reached at the end of the first injection peak.

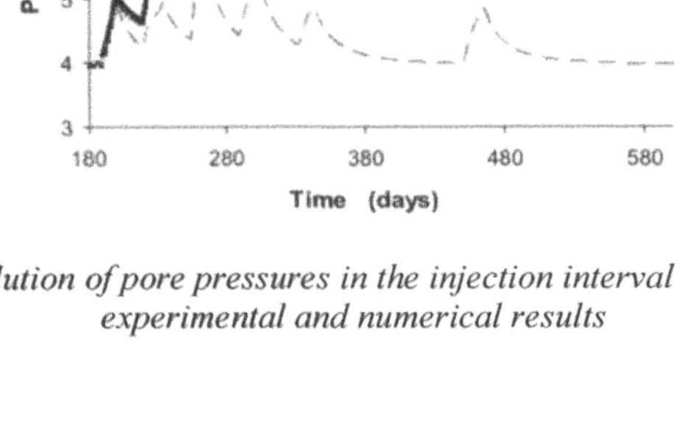

Figure 2.9. *Time evolution of pore pressures in the injection interval – comparisons between experimental and numerical results*

In the two other intervals located in the injection borehole on both sides of the injection interval (Figure 2.1), the modeling provides the same pore pressure evolution (Figure 2.10(a)), because the intervals have been fixed at the same distance from the injection interval. In this zone, the rock mass remains saturated and the variations of the pore pressures are due to water overpressures induced by the nitrogen injection. Nevertheless, a different behavior is observed experimentally in each interval, with higher pore pressures obtained in interval 3 that are well reproduced by the numerical results. In interval 2, the predictions are worse, but the orders of magnitude of the pore pressure variations are very low in both intervals. The difference between experimental and numerical results is thus not significant. Moreover, these sensors mainly detect the effects of fluid transfers along the interfaces between the packers and the rock mass. These effects are not reproduced numerically, because a perfect contact is assumed between these two components in our problem.

Three sensors have also been installed along the measuring borehole (Figure 2.1). As mentioned earlier, the rock mass remains saturated so far from the injection interval and only water overpressures are predicted by our model. The evolution of the pore pressures is well reproduced numerically, even though each injection peak is slightly overestimated (Figure 2.10(b)). The comparison between Figures 2.10(a) and (b) emphasizes the effect of permeability anisotropy, because pore pressure variations are higher along the measuring interval, while it is located further from the injection interval than intervals 2 and 3.

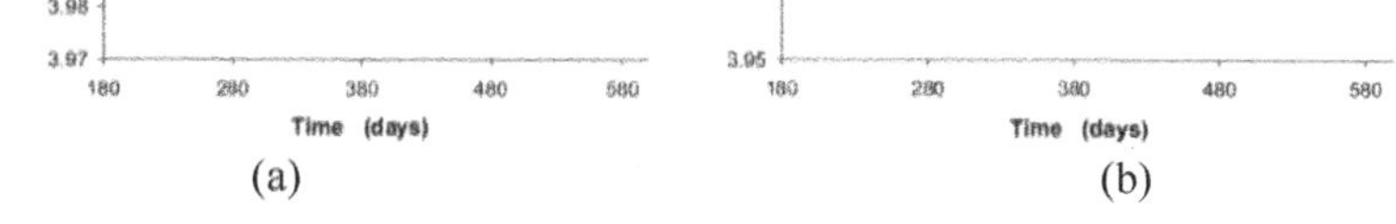

Figure 2.10. *Time evolution of pore pressures in a) the two intervals of the injection borehole and b) in the three intervals of the measuring borehole – comparisons between experimental and numerical results*

2.5. Discussion and conclusions

An *in situ* gas injection test has been performed in the Meuse/Haute-Marne Underground Research Laboratory in France. Nitrogen is injected at different flow rates from an injection interval, interrupted by shut-in phases. Different sensors provide the temporal evolution of pore pressures in the injection interval and in the rock mass. 1D and 3D hydraulic modeling of the problem have been performed. The analysis of the numerical results has shown the importance to know and take into account accurately each component of the experimental system, such as the volume and the initial conditions in the injection interval, the presence of a disturbed zone around the boreholes, and the rock mass characteristics. In particular, the way that water is removed from the injection interval or pushed in the rock mass influences strongly the analysis of the experimental observations. The numerical results are strongly dependent on the definition of the gas relative permeability in the quasi-saturated domain.

Moreover, even though the experiment has been initially designed to study the gas transfers in a potential host rock for radioactive waste disposal, the step-by-step 1D modeling has illustrated that the experimental response does not characterize the rock mass behavior at the beginning of the nitrogen injection. As confirmed by the gas flows profiles obtained from the best 1D modeling in a domain where the gaseous transfers are predominant (Figure 2.11), the first injection phase tests only the behavior of the injection interval, while the response of the second and the third peaks are influenced by the excavated damaged zone. It is only from the fourth peak that nitrogen reaches the undisturbed claystone and the pore pressure measurements are then influenced by its behavior.

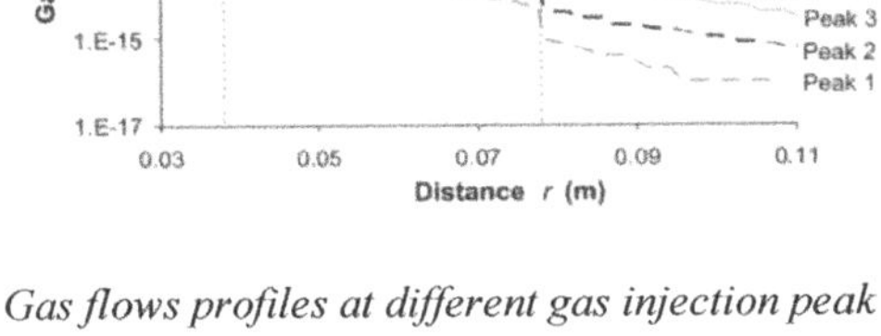

Figure 2.11. *Gas flows profiles at different gas injection peak – 1D modeling*

The following conclusions can be deduced from the experimental observations and from the different numerical modeling. The gas transfers are strongly dependent on the gas permeability in the quasi-saturated domain, where unfortunately only a few items of experimental data are available nowadays. The presence of an excavated damaged zone around the boreholes makes the gas entry in the rock mass easier and has to be considered in order to explain the experimental observations. In this disturbed domain, it is first the decrease of the gas entry pressure induced by the micro-fracturing rather than the increase of permeability that plays a major role on the gas transfers. The modeling has also confirmed previous experimental results illustrating that gas permeability in dried conditions is higher than water permeability in saturated conditions. Finally, such results show that a predictive model such as the two-phase flow approach is able to reproduce experimental observations in a large-scale system, as far as the injection flow rate and the gas pressures remain moderate. Taking into account the development of gas preferential pathways is certainly a crucial issue in the description of laboratory experiment, but seems to be neglected for this part of the test.

More generally, the PGZ1 experiment has shown that gas would remain confined mainly in the borehole disturbed zone. Even though gas penetrates in the sound claystone, the quantities remain low and are located near the injection interval with such gas injection conditions.

2.6. Bibliography

[AND 05] ANDRA, Projet HAVL – Dossier 2005. Référentiel site Meuse/Haute-Marne, Rapport Andra C.RP.ADS.04.0022, 2005.

[ARN 08] ARNEDO D., ALONSO E.E., OLIVELLA S., ROMERO E., "Gas injection tests on sand/bentonite mixtures in the laboratory. Experimental results and numerical modelling", *Physics and Chemistry of the Earth*, vol. 33, pp. S237–S247, 2008.

[BLÜ 07] BLÜMLING P., BERNIER F., LEBON P., DEREK MARTIN C., "The excavation damaged zone in clay formations time-dependent behaviour and influence on performance assessment", *Physics and Chemistry of the Earth*, vol. 32, 2007, p. 588–599.

[BOS 02] BOSSART P., MEIER P.M., MOERI A., TRICK T., MAYOR J.C., "Geological and hydraulic characterisation of the excavation disturbed zone in the Opalinus Clay of the Mont Terri Rock Laboratory", *Engineering Geology*, vol. 66, pp. 19–38, 2002.

[BOU 08a] BOULIN P.F., ANGULO-JARAMILLO R., DAIAN J.-F., TALANDIER J., BERNE P., "Experiments to estimate gas intrusion in Callovo-Oxfordian argillites", *Physics and Chemistry of the Earth*, vol. 33, pp. S225–S230, 2008a.

[BOU 08b] BOULIN P.F., ANGULO-JARAMILLO R., DAIAN J.-F., TALANDIER J., BERNE P., "Pore gas connectivity analysis in Callovo-Oxfordian argillite", *Applied Clay Science*, vol. 42, pp. 276–283, 2008b.

[CHA 12] CHARLIER R., COLLIN F., PARDOEN B., TALANDIER J., RADU J.-P., GERARD P., "An unsaturated hydro-mechanical modelling of two in-situ experiments in Callovo-Oxfordian argillite", *Engineering Geology*, in press, 2012.

[CUS 11] CUSS R.J., HARRINGTON J.F., NOY D.J., WIKMAN A., SELLIN P., "Large scale gas injection test (Lasgit): results from two gas injection tests", *Physics and Chemistry of the Earth*, vol. 36, pp. 1729–1742, 2011.

[DEL 02] DELAHAYE C.H., ALONSO E.E., "Soil heterogeneity and preferential paths for gas migration", *Engineering Geology*, vol. 64, pp. 251–271, 2002.

[DIS 07] DISTINGUIN M., LAVANCHY J.M., "Determination of the hydraulic properties of the Callovo-Oxfordian Argillites at the Bure Site: Synthesis of the results obtained in deep boreholes using several in-situ investigation techniques", *Physics and Chemistry of the Earth*, vol. 32, pp. 379–392, 2007.

[ESC 05] ESCOFFIER S., HOMAND F., GIRAUD A., HOTEIT N., SU K., "Under stress permeability determination of the Meuse/Haute-Marne mudstone", *Engineering Geology*, vol. 81, no. 3, pp. 329–340, 2005.

[GAL 00] GALLÉ C., "Gas breakthrough pressure in compacted Fo-Ca clay and interfacial gas overpressure in waste disposal context", *Applied Clay Science*, vol. 17, no. 1–2, pp. 85–97, 2000.

[GER 08] GERARD P., CHARLIER R., BARNICHON J.-D., SU K., SHAO J.-F., DUVEAU G., GIOT R., CHAVANT C., COLLIN F., "Numerical modelling of coupled mechanics and gas transfer around radioactive waste in long-term storage", *Journal of Theoretical and Applied Mechanics*, vol. 38, no. 1–2, pp. 25–44, 2008.

[GER 11] GERARD P., Impact des transferts de gaz sur le comportement poro-mécanique des matériaux argileux, PhD thesis University of Liège, Belgium, p. 284, 2011.

[GER 12] GERARD P., HARRINGTON J.F., CHARLIER R., COLLIN F., "Hydro-mechanical modelling of the development of preferential gas pathways in claystone", *Unsaturated Soils: Research and Applications*, MANCUSO C., JOMMI C., D'ONZA F. (Eds.), Springer, volume 2, pp. 175–180, 2012.

[GRI 94] GRINDROD P., IMPEY M.D., SADDIQUE S.N., TAKASE H., "Saturation and gas migration within clay buffers", *Proceedings of Conference on High Level Radioactive Waste Management*, Las Vegas, United States, April 1994.

[HAR 03] HARRINGTON J.F., HORSEMAN S.T., Gas migration in KBS-3 buffer bentonite. Sensitivity of test parameters to experimental boundary conditions, Technical Report SKB – TR-03-02, p. 57, 2003.

[HEI 02] HEITZ J.-F., HICHER P.-Y., "The mechanical behaviour of argillaceous rocks – Some questions from laboratory experiments", in HOTEIT N., SU K., TIJANI M., SHAO J.-F. (eds), *Hydromechanical and Thermohydromechanical Behaviour of Deep Argillaceous Rock-theory and Experiments*, Taylor & Francis, Paris, pp. 99–108, 2002.

[HIL 02] HILDENBRAND A., SCHLÖMER S., KROOS B.M., "Gas breakthrough experiments on fine-grained sedimentary rocks", *Geofluids*, vol. 2, pp. 3–23, 2002.

[HOM 04] HOMAND F., GIRAUD A., ESCOFFIER S., KORICHE A., HOXHA D., "Permeability determination of a deep argillite in saturated and partially saturated conditions", *International Journal of Heat and Mass Transfer*, vol. 47, pp. 3517–3531, 2004.

[HOR 99] HORSEMAN S.T., HARRINGTON J.F., SELLIN P., "Gas migration in clay barriers", *Engineering Geology*, vol. 54, pp. 139–149, 1999.

[LEV 10] LEVASSEUR S., CHARLIER R., FRIEG B., COLLIN F., "Hydro-mechanical modelling of the excavation damaged zone around an underground excavation at Mont Terri Rock Laboratory", *International Journal of Rock Mechanics and Mining Science,* vol. 47, no. 3, pp. 414–425, 2010.

[MAR 05] MARSCHALL P., HORSEMAN S.T., GIMMI T., "Characterisation of gas transport properties of the opalinus clay, a potential host rock formation for radioactive waste disposal", *Oil & Gas Science Technology*, vol. 60, no. 1, pp. 121–139, 2005.

[OLI 08] OLIVELLA S., ALONSO E.E., "Gas flow through clay barriers", *Géotechnique*, vol. 58, no. 3, pp. 157–168, 2008.

[ORT 97] ORTIZ L., VOLCKAERT G., DE CANNIÈRE P., PUT M., SEN M.A., HORSEMAN S.T., HARRINGTON J.F., IMPEY M.D., EINCHCOMB S., MEGAS: modelling and experiments on GAS migration in repository host rocks – Final report phase 2, European Report EUR17453EN, 1997.

[PHA 07] PHAM Q.T., VALES F., MALINSKY L., NGUYEN MINH D., GHARBI H., "Effects of desaturation-resaturation on mudstone", *Physics and Chemistry of the Earth*, vol. 32, pp. 646–655, 2007.

[SHA 08] SHAO H., SCHUSTER K., SÖNNKE J., BRÄUER V., "EDZ development in indurated clay formations – In situ borehole measurements and coupled HM modeling", *Physics and Chemistry of the Earth*, vol. 33, S388–S395, 2008.

[TSA 05] TSANG C.F., BERNIER F., DAVIS C., "Geohydromechanical processes in the Excavation Damaged Zone in crystalline rock, rock salt and indurated and plastic clays in the context of radioactive waste disposal", *Int. J. Rock Mech. Min. Sci.*, 2005;42:109–25.

[VAI 12] DE LA VAISSIÈRE R., TALANDIER J., "Gas entry pressure in Callovo-Oxfordian claystone: In-situ experiment PGZ1", in SKOCZYLAS F., DAVY C.A., AGOSTINI F., BURLION F. (eds), *Proceedings of Transfer 2012*, pp. 360–368, 2012.

[VAN 80] VAN GENUCHTEN M.T., "A closed-form equation for predicting the hydraulic conductivity of unsaturated soils", *Soil Science Society of America Journal,* vol. 44, pp. 892–898, 1980.

[VOL 95] VOLCKAERT G., ORTIZ L., DE CANNIÈRE P., PUT M., HORSEMAN S.T., HARRINGTON J.F., FIORAVANTE V., IMPEY M., MEGAS: modelling and experiments on gas migration in repository host rocks – phase 1 final report, European Commission Report EUR 16235 EN, 1995.

[XU 97a] XU K., DAIAN J.-F., QUENARD D., "Multiscale structures to describe porous media. Part I: Theoretical background and invasion by fluids", *Transport in Porous Media*, vol. 26, no. 1, pp. 51–73, 1997a.

[XU 97b] XU K., DAIAN J.-F., QUENARD D., "Multiscale structures to describe porous media. Part II: Transport properties and application to test materials", *Transport in Porous Media*, vol. 26, no. 3, pp. 319–338, 1997b.

[YAN 08] YANG D., Caractérisation par la mesure de perméabilité au gaz de l'endommagement mécanique et hydrique dans l'EDZ des argilites du Callovo-Oxfordien, PhD Thesis, Ecole des Mines de Paris, France, p. 208, 2008.

[ZHA 04] ZHANG C., ROTHFUCHS T., "Experimental study of the hydro-mechanical behaviour of the Callovo-Oxfordian argillite", *Applied Clay Science*, vol. 26, pp. 325–336, 2004.

Chapter 3

Upscaling Permeation Properties in Porous Materials from Pore Size Distributions

3.1. Introduction

The purpose of this chapter is to achieve a better understanding of the link between damage, failure and the transport properties of porous media. This is typically of utmost importance in the long-term assessment of the safety of CO_2 geological storage where the tightness of the caprock plays an important role. Due to the injection process and the transfer of acid gas into rock masses, loading–unloading cycles may occur. At the same time, chemomechanical effect may induce cracking, carbonation and possibly leaching of the host rock. These effects may have a direct influence on the tightness of the geological storage. Hence, addressing the relationship between the state of damage in rocks, their microstructure and their intrinsic permeability is a central issue.

For many years, authors have looked for predictive models of porous media permeability. In case of a perfect unimodal porous medium, Kozeny [KOZ 27] relates the permeability to the porosity, the tortuosity and an average pore size. The validity of this model was discussed later on by different authors (e.g. [CHA 03]). Katz and Thompson [KAT 86] predicted the permeability from the electrical conductivity of the porous material through a critical pores radius. Aït-Mokthar and co-workers [AÏT 99] proposed a model based on experimental data from a mercury intrusion porosimetry (MIP) and a pore size distribution (PSD) modeled with a statistical

Chapter written by Fadi KHADDOUR, David GRÉGOIRE and Gilles PIJAUDIER-CABOT .

log-normal distribution under the assumption that the pores are parallel cylinders of varying diameter. Later on, the same authors [AÏT 02] considered the porous medium as a three-dimensional cubic lattice with a bimodal distribution of pore sizes. An extension to a multi-modal PSD has been proposed by [AMI 05].

In these different models, the PSD experimentally measured is first idealized to estimate the permeability. Here, we propose a strategy that allows a direct estimation of the permeability from the measurements of the PSD. Hence, the relationship between the intrinsic material permeability and the mechanical damage, which induces micro-fracture or macro-fracture, is addressed through the evolution of the PSD due to mechanical damage. A model solely based on the PSD of the material and capable of predicting the material permeability is proposed. This model is used to predict the permeability of artificial rocks (mortar-type similar to tight rocks), which have experienced damage.

For illustration purposes, the present model is applied to the description of the variation of permeability due to electrical shock fracturing on a model material similar to tight rock, made of mortar. Here, electrical fracturing is based on the generation of a shock wave in water due to an electrical discharge between two electrodes. The shock wave is then transmitted to the solid material that is also immersed. Electrical shock fracturing is a prototype stimulation technique that is a potential alternative to classical hydraulic fracturing. This new technique has been recently proposed by [CHE 10], [MAU 10] and [CHE 12]. It is based on the fact that dynamic loads induce a fragmentation of the material distributed in the bulk, whereas quasi-static hydraulic fracturing induces a small number of cracks. Here, this technique is viewed as a means to generate damage, as distributed as possible, in a material in order to exhibit the relationship between the increase of an homogeneous damage and the increase of the intrinsic permeability of a representative volume of porous material. Results of experiments will be used to assess the accuracy of the prediction of the material permeability as a function of the PSD and its evolution under loading.

3.2. Assembly of parallel pores

3.2.1. *Presentation*

We consider here that the microstructure of the porous rock or model material consists of parallel and cylindrical pores of identical length (L_c) and different diameters (d_i) (see Figure 3.1).

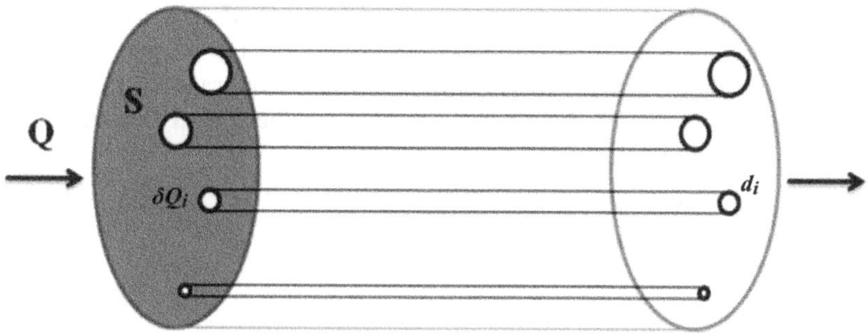

Figure 3.1. *Assembly of parallel pores*

3.2.2. *Permeability*

For a multi-modal porous material consisting of different pore sizes, the permeability is estimated by combining Darcy's law and Poiseuille's law, which represent the flow in the porous media at the macro and micro-scale, respectively (equations [3.1]–[3.9]). The pore size distribution is supposed to have been completely characterized (e.g. with the help of MIP). From the PSD (volume of pores of diameter d_i denoted as (V_{p_i}) vs. diameter (d_i)), the total length of each pore (L_i) is estimated through:

$$L_i = \frac{4V_{p_i}}{\pi d_i^2} \qquad [3.1]$$

A pore of length (L_i) is considered to cross the porous media as soon as its length reaches the critical length (L_c), which depends on the sample length through the tortuosity (T) (see equation [3.2]).

$$L_c = TL_e \quad \text{and} \quad L_e = \sqrt[3]{V_t} \qquad [3.2]$$

where L_e is the sample length and V_t is the total volume of porous media.

A volume V_t of porous media contains x_i pores of diameter d_i, where x_i is given as:

$$x_i = \left\lfloor \frac{L_i}{L_c} \right\rfloor \qquad [3.3]$$

where $\lfloor \bullet \rfloor$ is the floor function. The distribution of x_i is the quantity that is characteristic of the PSD and which we are going to use in the calculation of the permeability.

At the micro-scale, Poiseuille's law (equation [3.4], [POI 40]) gives the flow rate δQ_i in a pore of diameter d_i according to its length L_c, the pressure gradient ΔP and the dynamic fluid viscosity μ. By summation, the flow rate (Q_i) corresponding to all the pores of the same diameter d_i, as well as the global flow rate (Q) (see equation [3.5]), is obtained:

$$\delta Q_i = \frac{\pi}{128\mu}\frac{\Delta P}{L_c}d_i^4 = \frac{\pi}{128\mu T}\frac{\Delta P}{L_e}d_i^4 \quad [3.4]$$

$$Q = \sum_{i=1}^{N} Q_i = \sum_{i=1}^{N} x_i \delta Q_i = \sum_{i=1}^{N} \frac{\pi}{128\mu}\frac{\Delta P}{L_c} x_i d_i^4 = \frac{\pi}{128\mu}\frac{\Delta P}{TL_e}\sum_{i=1}^{N} x_i d_i^4 \quad [3.5]$$

At the macro-scale, Darcy's law (equation [3.6], [DAR 56]) relates the gradient of pressure to the flow rate (through the porous media apparent cross-section S and the sample length L_e). Therefore, we get an expression of the permeability K as a function of the PSD (see equation [3.7]):

$$Q = \frac{K}{\mu}\frac{\Delta P}{L_e}S \quad [3.6]$$

$$K = \left(\frac{\pi}{128S}\right)\left(\frac{1}{T}\right)\left(\sum_{i=1}^{N} x_i d_i^4\right) \quad [3.7]$$

The latter expression may be written as a function of the partial porosity ε_i, which corresponds to the partial volume V_{p_i} of pores of diameter d_i over the total volume V_t (see equations [3.8] and [3.9]):

$$\varepsilon_i = \frac{V_{p_i}}{V_t} = \frac{L_i \pi \frac{d_i^2}{4}}{V_t} = \frac{x_i T L_e \pi \frac{d_i^2}{4}}{SL_e} = \frac{x_i T \pi d_i^2}{4S} \quad [3.8]$$

$$K = \left(\frac{1}{32T^2}\right)\left(\sum_{i=1}^{N} \varepsilon_i d_i^2\right) \quad [3.9]$$

Note that this new expression of the permeability (equation [3.9]) corresponds to the classical Kozeny–Carman expression K^{KC} given in [CHA 03], in the case of an unimodal PSD (diameter d, porosity ε; see equation [3.10]):

$$K \equiv K^{KC} = \frac{\varepsilon}{32T^2}d^2 \quad [3.10]$$

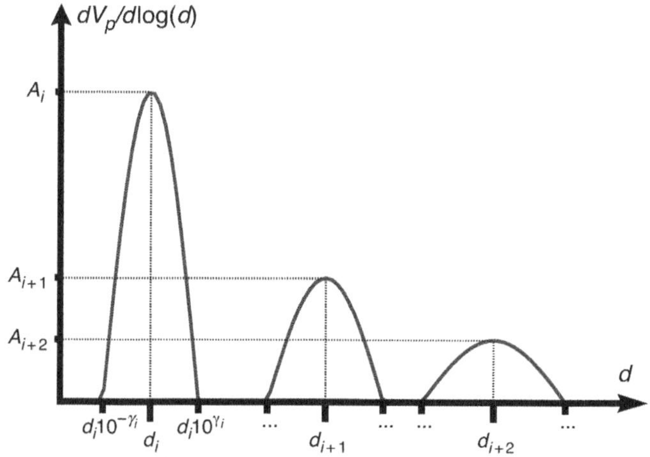

Figure 3.2. *Sinusoidal multi-peaks PSD (semi-log plotting)*

3.2.3. *Case of a sinusoidal multi-modal pore size distribution*

Let us apply this model to the case of a synthetic sinusoidal multi-modal PSD in semi-log axis (see Figure 3.2). In the case of a single sinusoidal peak (centered in d_i, amplitude A_i, period $4\gamma_i$, phase $\pi/2$), the PSD is given by:

$$\frac{dV_p}{d\log(d)} = A_i \sin\left(\frac{\pi \log \frac{d}{d_i}}{2\gamma_i} + \frac{\pi}{2}\right) \quad [3.11]$$

For N sinusoidal peaks with no overlapping, the porosity and the permeability are obtained analytically by equations [3.12] and [3.13], respectively.

$$\varepsilon^{\text{anal}} = \sum_{i=1}^{N} \int_{\log(d_i)-\gamma_i}^{\log(d_i)+\gamma_i} \frac{dV_{pi}}{V_t} = \frac{4}{\pi V_t} \sum_{i=1}^{N} A_i \gamma_i \quad [3.12]$$

$$K^{\text{anal}} = \frac{\varepsilon^{\text{anal}}}{32T^2} \sum_{i=1}^{N} \frac{\pi^2 (10^{2\gamma_i} + 10^{-2\gamma_i})}{2(16(\ln 10)^2 \gamma_i^2 + \pi^2)} d_i^2 \quad [3.13]$$

It follows from these relations that for a single-peak PSD (centered in d, period 4γ), the classical Kozeny–Carman (see equation [3.10]) is recovered with a correction corresponding to the PSD width:

$$K^{\text{single peak}} = K^{KC} \left(\frac{\pi^2 (10^{2\gamma_i} + 10^{-2\gamma_i})}{2(16(\ln 10)^2 \gamma_i^2 + \pi^2)} \right) \quad [3.14]$$

where K^{KC} is defined in equation [3.10].

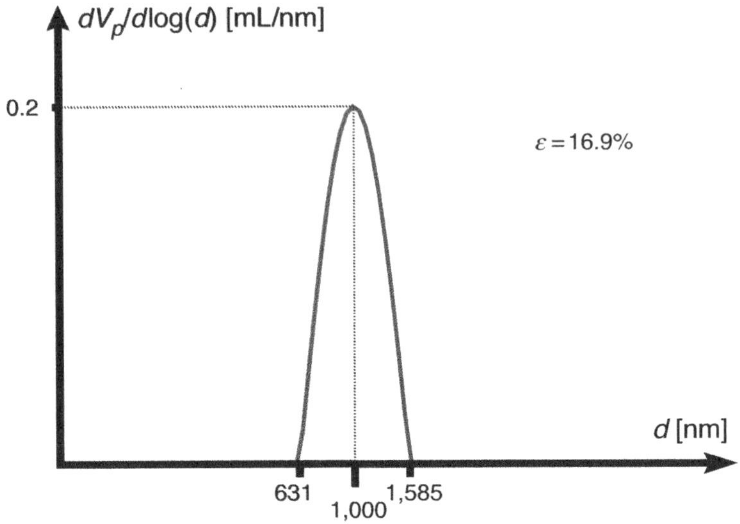

Figure 3.3. *Validation for a single-peak sinusoidal PSD (semi-log plotting)*

Figure 3.3 presents a typical sinusoidal single-peak PSD from which the permeability is computed. The obtained values of the permeability are:

- Model: $K \approx 5.713 \times 10^{-15}\,\text{m}^2$;
- Analytical: $K^{\text{anal}} \approx 5.718 \times 10^{-15}\,\text{m}^2$;
- Kozeny–Carman: $K^{KC} \approx 5.281 \times 10^{-15}\,\text{m}^2$.

The correction of the Kozeny–Carman formula, corresponding to the fact that the pores are not of a single size, is here equal to 1.082 and the permeability provided by the numerical model is very close to that of the analytical one (error: 0.1%), which simply underlines that the model has been implemented[1] in a correct way.

3.3. Mixed assembly of parallel and series pores

3.3.1. *Presentation*

The main drawback of the previous model based on a simple assembly of parallel pores is that a pore with a length lower than the critical length has no contribution to the permeability, and the fractional part of their ratio as well. An enhanced model can

1 Practically the computations are performed with Scilab (www.scilab.org).

be devised where larger pores are connected in series with smaller pores following Figure 3.4.

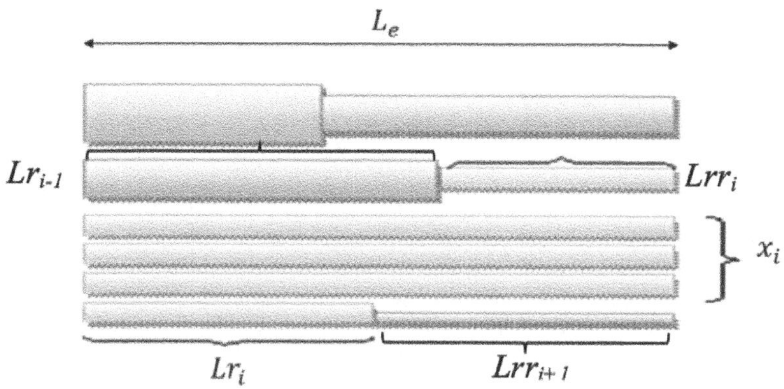

Figure 3.4. *Mixed assembly of pores in parallel and series*

We consider here that the microstructure consists of parallel and cylindrical pores of identical length ($L_c = TL_e$) and different diameters (d_i). For each pore of diameter d_i and volume V_{pi}, its length is given by $L_i = 4V_{pi}/\pi d_i^2$. We denote L_{rri} the pore length connected to the previous pore of diameter d_{i-1}, L_{ri} the pore length connected to the next pore of diameter d_{i+1} and x_i the number of parallel pores that have a given length equal to L_c. Therefore, the total pore length for a given pore diameter d_i is given by:

$$L_i = L_{ri} + L_{rri} + x_i L_c \qquad [3.15]$$

Practically, Figure 3.5 presents the flowchart that leads to the mixed assembly determination meaning the estimation of, respectively, L_{rri}, x_i and L_{ri} ($\lfloor \bullet \rfloor$ is still the floor function).

3.3.2. *Permeability*

For a multi-modal porous material consisting of different pore sizes, the permeability is still estimated by combining Darcy's law and Poiseuille's law but at the micro-scale, the flow rate (Q) given by the Poiseuille's equation is decomposed into two terms depending if the pores are assembled in parallel (Q_p) or in series (Q_a):

$$Q = Q^p + Q^a = \sum_{i=1}^{N} Q_i^p + \sum_{i=1}^{N} Q_i^a = \underbrace{\sum_{i=1}^{N} \frac{\pi}{128\mu} \frac{\Delta P}{L_c} x_i d_i^4}_{\text{following equation [3.5]}} + \sum_{i=1}^{N} Q_i^a \qquad [3.16]$$

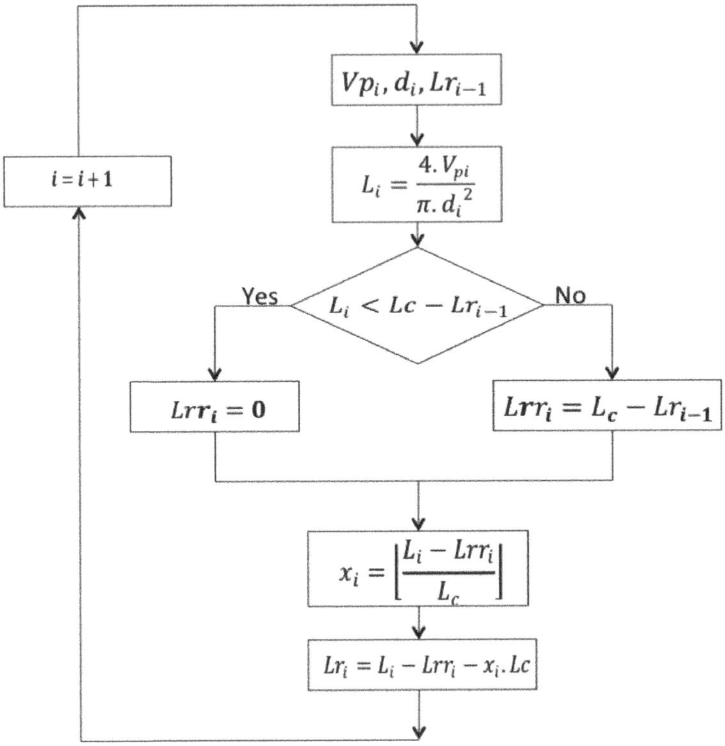

Figure 3.5. *Flowchart for mixed assembly determination*

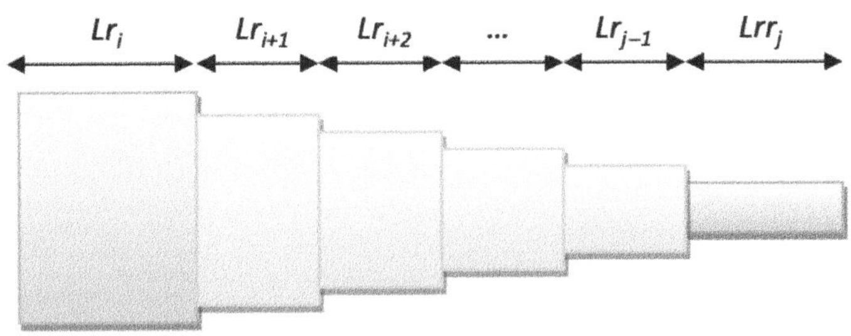

Figure 3.6. *Assembly of pores in series from pore diameter d_i to d_j*

The term corresponding to the parallel assembling is still given by equation [3.5], whereas the term corresponding to a series assembling has to be derived. Figure 3.6 presents a typical assembly of pores in series. Let us denote the first pore diameter by

d_j such as $L_{rrj} \neq 0$, the flow corresponding to an assembly of pores in series is then given by:

$$Q_i^a = \frac{\pi}{128\mu}\Delta P \frac{1}{\sum_{k=i}^{j-1}\left(\frac{L_{rk}}{d_k^4} + \frac{L_{rrj}}{dj^4}\right)} \quad \text{where } j = \min_{k\in \mathrm{N}}\{k \geq i+1 | L_{rrk} \neq 0\} \qquad [3.17]$$

Therefore, the overall permeability, estimated with reference to Darcy's law (see equation [3.6]), is given as:

$$K = \frac{\mu Q L_e}{\Delta P S} = \frac{\mu L_e}{\Delta P S}(Q^p + Q^a) = \frac{\pi L_c}{128 S T}\sum_{i=1}^{N}\left(\frac{x_i d_i^4}{L_c} + \frac{1}{\sum_{k=i}^{j-1}\left(\frac{L_{rk}}{d_k^4} + \frac{L_{rrj}}{dj^4}\right)}\right)$$
$$\text{where } \forall i \in [1, N],\ j = \min_{k\in \mathrm{N}}\{k \geq i+1 | L_{rrk} \neq 0\} \qquad [3.18]$$

3.4. Comparisons with experimental results

3.4.1. *Electrical fracturing tests*

Experimental tests have been carried out by [CHE 10] and [MAU 10] to study the evolution of micro-cracking and the inherent change of permeation properties during dynamic electrical shock fracturing in natural rocks and mortar. In this section, we use these data for illustrative purposes and we will compare some of the experimental results with our model predictions. The details of these experiments, which are not of interest considering the scope of the present volume, are presented in [CHE 10] and [MAU 10].

Cylindrical mortar specimens and point–point electrode system are immersed in a tank filled with tap water at room temperature. A compression shock wave is generated by an electrical discharge above the specimen. This compression wave then propagates into the water and inside the specimen. The pressure wave amplitude is constant (90 MPa) and is controlled by the variation of the injected electrical energy and by the distance between the electrodes and the specimen surface. After each run of shock waves, the axial vertical permeability to gas is measured. Figure 3.7 shows the experimental setup and Table 3.1 shows the number of shocks and the corresponding values of permeability for four specimens tested by [CHE 10]. We shall consider here the tests where repeated loads of the same amplitude have been applied on the same specimen. Figure 3.8 shows the influence of the number of electrical shocks on intrinsic permeability for all specimens tested by [CHE 10]. It is observed that the intrinsic permeability of the specimens increases with the number of shocks. It is also observed that specimens subjected to more than eight electrical shocks present severe damage, micro-cracks and even macro-cracks (see [MAU 10] for details). At this stage, the distribution of damage within the specimen is no longer homogeneous and the calculation of the permeability from the PSD, which assumes that the pore distribution is homogeneous over the specimen, cannot be applied directly.

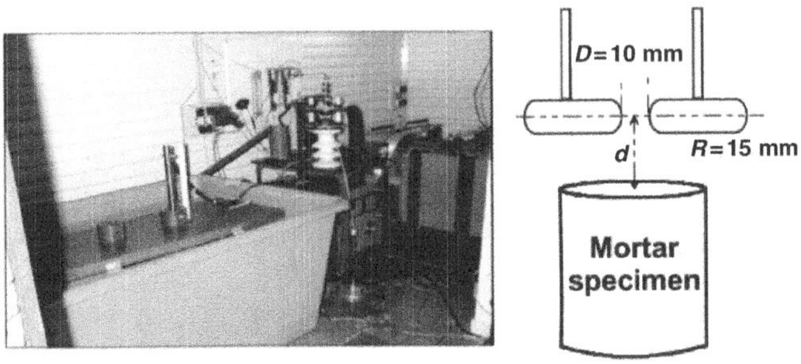

Figure 3.7. *Experimental setup (reproduced from [MAU 10])*

Specimen	$B2$	$H14$	IXB	$8B$
Number of shock	0	3	6	10
Permeablity (m^2)	3×10^{-17}	8.5×10^{-17}	1.95×10^{-16}	3.08×10^{-17}

Table 3.1. *Number of electrical shock and corresponding permeability (reproduced from [CHE 10])*

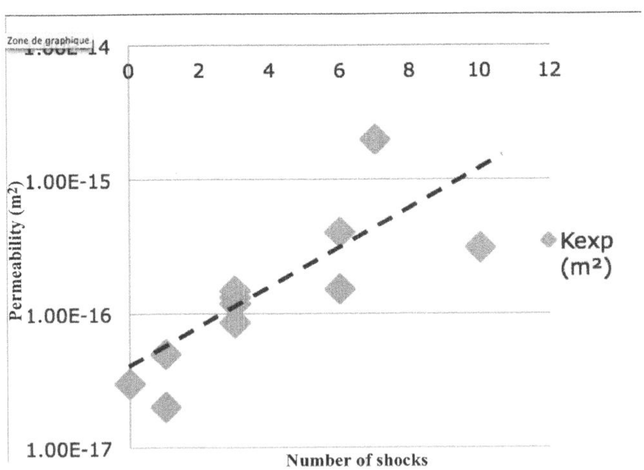

Figure 3.8. *Influence of the number of electrical shocks on intrinsic permeability (reproduced from [MAU 10])*

3.4.2. *Measurement of the pore size distribution*

MIP is a classical technique from which the PSD of a porous material can be obtained rather consistently if the pore size ranges between tens of nanometers and up to hundreds of micrometers (typically from 20 nm to 200 μm). Each specimen submitted to electrical fracturing then has to be subjected to mercury intrusion to characterize the PSD and its evolution with the successive application of dynamic loads. The sample preparation consists of drilling small cylinder (20 mm) from the electrical shock specimens. The samples are then submitted to degas conditions up to 24 h (low vacuum, 90°C). Three samples are extracted from each specimen are presented in Table 3.1. The PSD are then averaged.

Figures 3.9 and 3.10 show the evolution of PSD for specimens subjected to 0, 3, 6, and 10 shocks. As the number of shock increases, the microstructure evolves. From 0 to 6 shocks (Figure 3.9), small pores grow (pore radius $\approx$ 50 nm) whereas large pores shrink (pore radius $\approx$ 250 nm). Indeed, the compressive wave generated by the electrical shock induces compaction with a relative decrease of the number of large pores and an increase of the number of small pores.

For specimens subjected to 8 and 10 shocks (Figure 3.10), the PSD does not follow the same evolution. The specimens subjected to more than eight electrical shocks present severe damage, micro-cracks and possibly macro-cracks. The dynamic load now tends to increase the number of large pores. The compaction observed previously seems to be reversible and the evolution of the PSD is a simple translation toward the larger pores (recall that the PSD for a large amount of shocks may not be representative as macro-cracking is localized in the specimen).

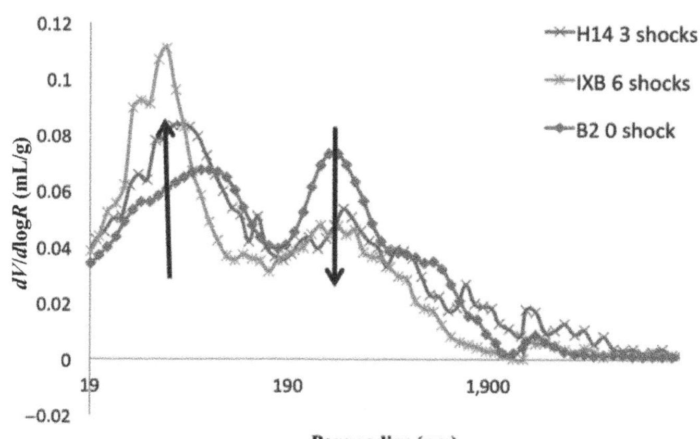

Figure 3.9. *Influence of the number of electrical shocks on PSD measurements (from 0 to 6 shocks)*

3.4.3. *Model capabilities to predict permeability and comparisons with experiments*

From the different PSDs presented in Figures 3.9 and 3.10, permeabilities have been estimated through the mixed assembly of parallel and series pores model.

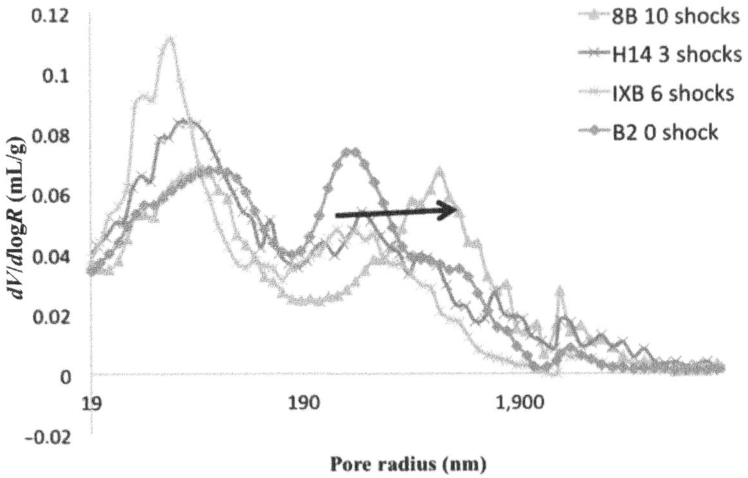

Figure 3.10. *Influence of the number of electrical shocks on PSD measurements (from 0 to 10 shocks)*

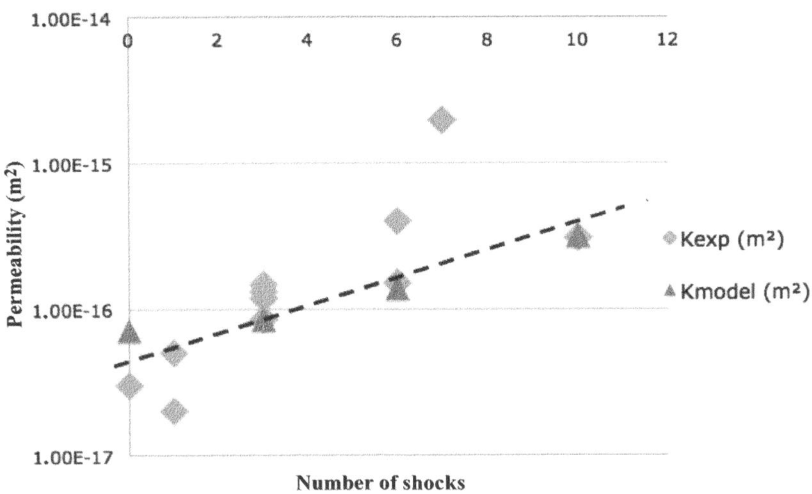

Figure 3.11. *Experimental and model comparisons of the permeability evolution with the number of electrical shocks*

Figure 3.11 shows the evolution of permeability with the number of electrical shocks and the comparisons with the experimental results. A good concordance is obtained even if the range of variation of the PSD does not allow a comprehensive comparison.

3.5. Conclusions

A new model that estimates the overall permeability of a porous material directly from PSD has been presented. A parallel assembly of pores and a mixed parallel–series assembly of pores have been presented. In both cases, the permeability is estimated by combining Darcy's law and Poiseuille's law, which represent the flow in the porous media at the macro-scale and at the micro-scale, respectively. In this respect, the present model falls in the category of upscaling techniques for permeation properties in porous solids. This model represents an improvement to classical Kozeny–Carman model types where only one pore diameter is considered.

First, the model has been compared to the classical Kozeny–Carman model and validated using a synthetic sinusoidal PSD. Second, the model predictions have been compared with the experimental permeability measurement performed during dynamic electrical shock fracturing tests. It has been shown that the model predictions of permeabilities are very close to the experimental permeabilities corresponding to the same specimens.

The main difficulty for subsequent challenging comparisons over a wide range of variation of permeability remains in the fact that the permeability has been measured on specimens of a different size compared to those used for mercury intrusion. If the distribution of micro-cracking is not homogeneous over the specimen, and therefore the PSD is not the same over the specimen, then some spurious effects may appear on the estimation of permeability and on the relationship between damage and permeation properties. One needs either to achieve homogeneously distributed evolution of micro-craking over the specimen tested – which is quite difficult as strain and damage will certainly localize in the course of failure – or to use the same size and geometry of specimens in all the types of experiments (mechanical, permeation and MIP). It is this second track that is currently being followed.

3.6. Acknowledgments

This work was sponsored by the ERC advanced grant Failflow (27769). This financial support is gratefully acknowledged. F. Khaddour is grateful to the Syrian Ministry of Higher Education for its support through Grant no. 13153/4/W to visit the University of Pau and Pays de l'Adour as a doctorat student.

3.7. Bibliography

[AÏT 99] AÏT-MOKHTAR A., AMIRI O., SAMMARTINO S., "Analytic modelling and experimental study of the porosity and permeability of a porous medium-application to cement mortars and granitic rock.pdf", *Magazine of Concrete Research*, vol. 51, no. 6, pp. 391–396, 1999.

[AÏT 02] AÏT-MOKHTAR A., AMIRI O., DUMARGUE P., SAMMARTINO S., "A new model to calculate water permeability of cement materials from MIP results", *Magazine of Concrete Research*, vol. 14, no. 2, pp. 43–49, 2002.

[AMI 05] AMIRI O., AÏT-MOKHTAR A., SARHANI M., "Tri-dimensional modelling of cementitious materials permeability from multimodal pore size distribution obtained by mercury intrusion porosimetry tests", *Advances*, vol. 17, no. 1, pp. 39–45, 2005.

[CHA 03] CHAPUIS R. P., AUBERTIN M., Predicting the coefficient permeability of soils using the Kozeny–Carman equation, Report no. January, Département des génies civil, géologique et des mines, École Polytechnique de Montréal., 2003.

[CHE 10] CHEN W., Fracturation électrique des géomatériaux. Etude de l'endommagement et de la perméabilité, PhD thesis, University of Pau and Pays de l'Adour, 2010.

[CHE 12] CHEN W., MAUREL O., REESS T., DE FERRON A. S., LA BORDERIE C., PIJAUDIER-CABOT G., REY-BETHBEDER F., JACQUES A., "Experimental study on an alternative oil stimulation technique for tight gas reservoirs based on dynamic shock waves generated by pulsed arc electrohydraulic discharges", *Journal of Petroleum Science and Engineering*, vol. 88–89, pp. 67–74, 2012.

[DAR 56] DARCY H., *Les fontaines publiques de la ville de Dijon*, Dalmont, 1856.

[KAT 86] KATZ A., THOMPSON A., "Quantitative prediction of permeability in porous rock", *Physical Review. B, Condensed Matter*, vol. 34, no. 11, pp. 8179–8181, American Physical Society, 1986.

[KOZ 27] KOZENY J., "Ueber kapillare Leitung des Wassers im Boden", *Sitzungsberichte der Akademie der Wissenschaften, Wien*, vol. 136, no. 2a, p. 271, 1927.

[MAU 10] MAUREL O., REESS T., MATALLAH M., DE FERRON A., CHEN W., LA BORDERIE C., PIJAUDIER-CABOT G., JACQUES A., REY-BETHBEDER F., "Electrohydraulic shock wave generation as a means to increase intrinsic permeability of mortar", *Cement and Concrete Research*, vol. 40, no. 12, pp. 1631–1638, Elsevier Ltd, 2010.

[POI 40] POISEUILLE J., "Recherches expérimentales sur le mouvement des liquides dans les tubes de très-petits diamètres", *Comptes rendus hebdomadaires des séances de l'Académie des sciences*, vol. T11, no. 1, pp. 1041–1048, 1840.

PART 2

Fracture, Deformation and Coupled Effects

Chapter 4

A Non-Local Damage Model for Heterogeneous Rocks – Application to Rock Fracturing Evaluation Under Gas Injection Conditions

Carbon capture and storage (CCS) in deep geological formations are considered as one of the solutions for reducing the negative effects of greenhouse gas concentration. Demonstration of the safety and performance of the storage is a key factor in view of the industrial development of this technology. Geological layers, as the host formation of storage, contain different heterogeneities. The main goal of this chapter is to introduce a damage model taking into account the effects of the rock heterogeneities on its fracturing. A probabilistic damage model is used to study possible cracking of the rock due to the effective stress changes. Two different thresholds are considered for crack initiation and propagation. A Weibull model is used to account for the stochastic nature of crack initiation(s) and then a fracture mechanics-based threshold is considered to account for crack propagation. The formulation is integrated in a finite element package as a non-local damage model. Both damage thresholds are checked using a "regularized" stress field to avoid mesh dependence and localization phenomena. A sequential approach is proposed for hydromechanical analyses. Capacities of the developed model to evaluate the reliability of the caprock are illustrated through a typical fluid injection example.

Chapter written by Darius M. SEYEDI, Nicolas GUY, Serigne SY, Sylvie GRANET and François HILD.

The effects of rock heterogeneity on the sustainable injection pressure and possible patterns of induced damage are investigated through a set of parametric calculations.

4.1. Introduction

All underground engineering works have to deal with heterogeneous rock mass (e.g. nuclear waste disposal in geological media, deep underground injection, geothermal energy extraction, enhanced recovery from oil and gas reservoirs and underground storage of natural gas). More recently, CCS has been suggested as an important potential method to reduce the emission of greenhouse gases into the atmosphere. Deep saline aquifers are of main interest for geological storage as they can provide important storage capacities. Demonstration of the safety and integrity of CCS projects is a key factor for industrial deployment of this technology. CO_2 injection may induce various geochemical, thermo-hydraulic and geomechanical phenomena in different spatial and temporal scales (e.g. [PRU 02, GAU 08, RUT 08, ROH 10]). The effects of these phenomena on the performance and integrity of the storage site must be studied. A scenario-based safety approach is always used for defining safety criteria for CO_2 geological storage (e.g. [BOU 09]). Different geomechanical aspects arise for evaluating risk scenarios, such as caprock integrity, fault reactivation, surface uplift or subsidence and well integrity.

Evaluation of the sustainable injection pressure is one of the main parts of the safety study of a CO_2 storage site. Deep well injection changes the stress state in the geological formation. A change in the stress field may affect the hydraulic response of the medium through induced damage in the rock mass. The main goal of the present work is to introduce a hydromechanical damage model that takes into account the effect of mechanical heterogeneities on the fracturing probability due to gas injection operations.

CO_2 injection in a deep aquifer induces a pressure buildup in the geological reservoir and changes the total and effective stresses in the reservoir and adjacent geological formations (e.g. [RUT 08, VID 09, ROH 10]). This local stress modification may lead to damage for high-pressure buildups. Two main mechanical degradation mechanisms can be distinguished, namely, rock fracturing induced by tensile stresses and shear slip of the existing fractures, or rock plasticity induced by shear stresses (e.g. [RUT 08, SOL 09, ROH 10]). The prevalent degradation mechanism to study depends on the injection process, the initial stress regime and on site and rock properties. Guy *et al.* [GUY 10] have studied the influence of the initial stress state on the possible fracture mechanisms through a set of parametric hydromechanical calculations. Sustainable injection rate maps are then proposed providing a first-order estimate depending on initial stress states.

In the following, the mechanical damage model is first presented. A sequential approach is then introduced for hydromechanical coupling. The proposed model is finally applied to a typical gas injection example. The possibility of damage initiation in the caprock due to the pressure buildup on the reservoir–caprock interface is investigated.

4.2. A probabilistic non-local model for rock fracturing

The failure of a rock sample has a random character. The failure stress is scattered and the average level decreases with the volume of the sample (e.g. [BIE 75, BAE 81]). The effect of this scatter on the mass stability and its capacity to store gas must be taken into account. The rock mass heterogeneity is described by the presence of defects with a random distribution. The scatter of failure stress for rocks can be explained by the presence of microfractures (i.e. initial defects) that are at the origin of crack initiation, causing its failure. A probabilistic model based on a Poisson point distribution of microfractures is used to describe this random character by relating the material microstructure and its macroscopic behavior.

In this chapter, a probabilistic non-local model [GUY 12] is used to study possible cracking of the rock mass due the effective stress changes. Two different thresholds are considered for crack initiation and propagation. A Weibull model is used to account for the stochastic nature of crack initiation(s) and then a fracture mechanics-based threshold is considered to model crack propagation. Both thresholds are probed by using a "regularized" stress field to avoid mesh dependence and localization phenomena. The regularization operation is performed over a characteristic length l_c that must be long enough with respect to the size of the element in the zone of interest. The regularization operator reads

$$\bar{\boldsymbol{\sigma}} - l_c^2 \Delta \bar{\boldsymbol{\sigma}} = \boldsymbol{\sigma}, \text{ with } (\nabla \bar{\boldsymbol{\sigma}}) \cdot \boldsymbol{n} = 0 \quad [4.1]$$

where $\bar{\boldsymbol{\sigma}}$ is the regularized stress tensor, $\boldsymbol{\sigma}$ Cauchy's stress tensor, l_c a characteristic length and $\boldsymbol{n}$ the normal vector to a surface where the natural boundary conditions are considered. The procedure used is similar to the implicit gradient enhanced scheme proposed by Peerling *et al.* [PEE 98, PEE 01].

In this setting, the initiation of a new macro-crack follows a Weibull model. An initiation probability $P_i\,(el) \in [0;1]$ calculated from a uniform distribution is assigned to each subdomain (i.e. each element). An initiation stress is then obtained for each subdomain from an inverse Weibull law

$$S_i(el) = \frac{\sigma_0}{(\lambda_0 Z_{el})^{\frac{1}{m}}} \left[-ln(1 - P_i(el)) \right]^{\frac{1}{m}} \qquad [4.2]$$

where S_i is the initiation threshold, Z_{el} the size of the considered element, m the Weibull modulus and σ_0^m / λ_0 a scale parameter. A crack will initiate in the considered element if the regularized maximum principal stress reaches S_i. The initiated macro-cracks are assumed to be perpendicular to the regularized maximum principal stress direction.

Considering the two-dimensional case of a crack in an elastic medium submitted to a far field loading, the asymptotic Westergaard's solution [KAN 85] gives a good approximation of the stress field around a crack tip. Considering equation [4.1] as a non-homogeneous Helmholtz's equation, the proposed operator provides a regularized stress at the crack tip corresponding to Westergaard's asymptotic solution [GUY 12]. The present procedure allows the stress intensity factor to be calculated without any mesh refinement for a propagating crack. Therefore, a natural transition from a damage mechanics-based model for crack initiation to a fracture mechanics-based model for crack propagation is provided. In addition, both models use the same variable, that is the regularized stress. In this framework, the crack propagation threshold reads [GUY 12]

$$S_p(el) = \frac{6\Gamma^2\left(\frac{3}{4}\right)}{5\pi} \frac{K_c}{\sqrt{\pi l_c}} \qquad [4.3]$$

where K_c is the fracture toughness of the rock and Γ the gamma function. In this framework, cracks grow perpendicular to the maximum regularized principal stress direction. For a mode I crack, the above-mentioned criterion tends to a fracture mechanics threshold for an open crack when the characteristic length l_c becomes small compared to the macroscopic scale. It is to be noted that the introduced criterion accounts for mode mixity as the regularized stress reads

$$K_c = \frac{\left[5K_I + \sqrt{K_I^2 + 16K_{II}^2}\right]}{6} \approx K_I + \frac{2}{3}K_{II} \qquad [4.4]$$

where K_{II} stands for the mode II stress intensity factor.

A perfectly brittle behavior of an isotropic material is considered as a local damage law. In this setting, the Helmholtz's state potential reads

$$\rho\psi_e = \frac{1}{2}(1-d)\epsilon : \boldsymbol{C} : \epsilon \qquad [4.5]$$

where $\boldsymbol{C}$ is the elasticity tensor of undamaged material, ρ the mass density, d the damage variable and $\boldsymbol{\varepsilon}$ the strain tensor. The elasticity law thus reads

$$\boldsymbol{\sigma} = (1-d)\boldsymbol{C} : \epsilon \qquad [4.6]$$

The thermodynamic force Y associated with the damage variable d is defined as

$$Y = -\rho\frac{\partial\psi_e}{\partial d} = \frac{1}{2}\epsilon : \boldsymbol{C} : \epsilon \qquad [4.7]$$

Damage growth is written as

$$d = H\left(\langle\bar{\sigma}_I - S_i\rangle + \langle\bar{\sigma}_I - S_p\rangle\right) \text{ when } \dot{d} \geq 0 \qquad [4.8]$$

where H denotes Heaviside's step function, $\bar{\sigma}_I$ the maximum regularized stress and S_i and S_p the initiation and propagation thresholds, respectively.

The crack initiation criterion is introduced considering that a crack initiation stress is associated with each initial defect. The initial defects greater than $2 \times l_c$ cannot be considered in this way. Let us consider a domain with a critical defect of half-length a. Assuming that the characteristic length is greater than a leads to

$$S_i = \frac{K_c}{\sqrt{\pi a}} \geq \frac{K_c}{\sqrt{\pi l_c}} = \frac{5\pi}{6\Gamma^2\left(\frac{3}{4}\right)} S_p > S_p \qquad [4.9]$$

A characteristic length greater than the size of the largest initial defect provides a crack initiation threshold greater than the propagation threshold. The characteristic length corresponds to a microstructural parameter that is the size of the largest initial crack modeled as a defect. Initial cracks of size larger than $2 \times l_c$ can be modeled as a set of completely damaged elements.

4.3. Hydromechanical coupling scheme

Extensive works can be found in the literature dealing with coupled fluid flow and mechanics in porous media (e.g. [BIO 41, PAR 83, BOR 95, SCH 04]). Different coupling strategies can be used to this end. These strategies can be split into four major classes [SET 01, DEA 06, KIM 10]:

1) Fully coupled approach. Governing equations of flow and mechanics are solved simultaneously during each time step. An iterative scheme, typically a Newton–Raphson method, is used to obtain the converged solution.

2) Sequential scheme. Flow and mechanical problems are solved sequentially. Intermediate information from the first solved problem, that is flow or mechanical, is linked to the problem that is solved second. An iterative scheme is used to obtain a converged solution.

3) Staggered or single-pass sequential approach. It is a special case of the sequential scheme, where only one iteration is performed (e.g. [PAR 83]).

4) Loosely coupled scheme. The coupling between flow and mechanical problems is not performed on all the time steps.

Sequential schemes are commonly developed and used because fully coupled schemes could be very expensive from a computational point of view. In addition, in a sequential scheme, two separate simulators, each specific to the problem of interest, can be used.

Numerical modeling of CO_2 injection integrity and performance must also consider coupling of various phenomena, namely geochemistry, geomechanics, fluid and heat flow, and transport. To the authors' knowledge there is no single simulator that can perform large-scale calculations considering all aforementioned phenomena. Sequential coupling of specified codes is always used to this end. Rutqvist *et al.* [RUT 02a] developed a sequential approach coupling THOUGH2 and FLAC3D for the analysis of coupled multiphase fluid flow, heat transfer and deformation in porous rocks. Rutqvist and Tsang [RUT 02b] then applied this simulator to study the caprock hydromechanical changes associated with CO_2 injection into an aquifer. Vidal-Gilbert *et al.* [VID 09] used a "one way" coupling of reservoir and geomechanical analysis, where pressures obtained from reservoir simulations were integrated as input for a geomechanical model. In a more recent work, Rohmer and Seyedi [ROH 10] developed a sequential coupling of THOUGH2 and Code_Aster for large-scale hydromechanical models of a storage complex. TOUGH2 [PRU 99], a THC code and Code_Aster [EDF 08], a general thermo-hydromechanical code, are linked using sequential execution and data transfer.

In the present work, hydraulic and mechanical calculations are performed sequentially using different modules of Code_Aster. Figure 4.1 shows a schematic view of the scheme.

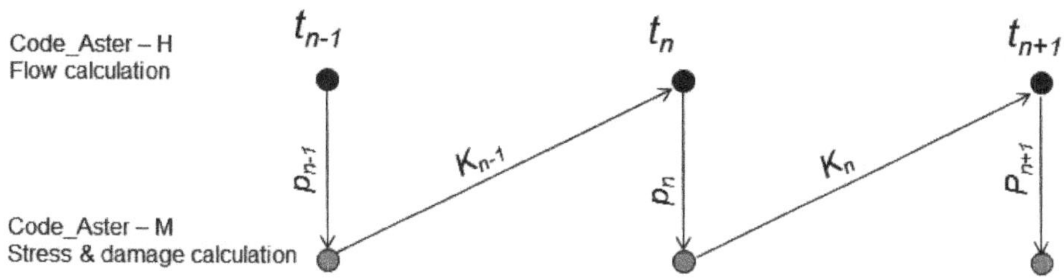

Figure 4.1. *Schematic representation of sequentially linking hydraulic and mechanical calculations with p_n being the pressure, K_n permeability and t_n time at the end of the nth time step*

At step *n*, the pore pressure *p* in the whole geological medium is calculated by the hydraulic simulator. The pressure is then transferred to the mechanical code as a body force. In this manner, the effect of the pressure change on the effective stresses is obtained considering Biot's coefficient.

In this chapter, only the overpressure induced by the gas injection is considered. The whole domain is considered as saturated. The following governing equations are considered in the hydraulic calculation. The conservation equation reads

$$\dot{r}_m + div(\boldsymbol{M}) = 0 \qquad [4.10]$$

where $r = r_m + r_0$ is the average global mass, r_0 the initial average global mass, r_m the incoming mass, and $\boldsymbol{M}$ the mass flux. The fluid compressibility is described by

$$\frac{d\rho_l}{\rho_l} = \frac{dp}{K_w} \qquad [4.11]$$

where ρ_l is the liquid density, and K_w its bulk modulus. Fluid diffusion is expressed as

$$\frac{\boldsymbol{M}}{\rho_l} = \lambda(-\nabla(p) + \rho_l \boldsymbol{g}) \qquad [4.12]$$

where $\lambda = K_{int}/\mu_l$ is the fluid diffusivity, K_{int} the intrinsic permeability, and μ_l the dynamic viscosity of the fluid.

The mechanical problem follows the resolution of the flow problem using an appropriate time discretization scheme. In this setting, the volumetric strain term of the porosity change (equation [4.14]) is evaluated explicitly. The flow problem is solved while the strain remains constant in the whole model. In the same manner, the pressure is prescribed during the resolution of the mechanical problem as the pressure at t_{n+1} is determined from the previous flow problem. The pressure corrections are considered as "loads" for solving the mechanical problem [SET 98]. The momentum equation thus reads

$$div\boldsymbol{\sigma}' - b(\nabla.p) + r\boldsymbol{g} = 0 \qquad [4.13]$$

where $\boldsymbol{\sigma}'$ is the effective stress tensor, b the Biot's coefficient, and p the pressure. For an elastic behavior, the porosity φ will be updated using the following expression:

$$d\varphi = (b-\varphi)\left(d\varepsilon_v + \frac{dp}{K_s}\right) \qquad [4.14]$$

where ε_v is the volumetric strain, and K_s the compressibility of the solid phase. Effective stress changes are thus calculated everywhere at each time step.

4.4. Application example and results

It has been shown that the reservoir–caprock interface constitutes a critical part of the storage complex where damage may occur due to the pressure buildup in the reservoir for high values of the injection pressure (e.g. [ROH 10]). The developed model is used to evaluate the damage probability of a hypothetical caprock. To this end, a part of a caprock of 1,000 m in length and located between 900 and 1,000 m depth is considered. An extensional stress regime, that is $|\sigma_h| < |\sigma_v|$, is considered, which is a situation where vertical fractures may be initiated in the caprock [GUY 10]. The effect of the injection procedure is idealized by a gradual increase in the pore pressure on the caprock–reservoir interface. Three zones are considered. In the middle zone (a segment of 100 m) of the bottom line of the caprock (see Figure 4.2), the pressure increases from the initial hydrostatic pressure p_0 to $3p_0$ during 5 years. During the same period, on two 200-m-long segments on both sides of the middle part, the pressure increases from p_0 to $2p_0$. Finally, the pressure remains constant elsewhere on the bottom boundary of the caprock. It is assumed that the pressure remains constant on the upper boundary of the caprock. Lateral boundaries of the caprock are considered as impermeable (i.e. flux is set to zero). Vertical displacements on the bottom and horizontal displacements on the lateral boundaries are set to zero. Table 4.1 summarizes the main hydromechanical parameters used in the calculation. No coupling from the mechanical calculation on the flow problem is considered.

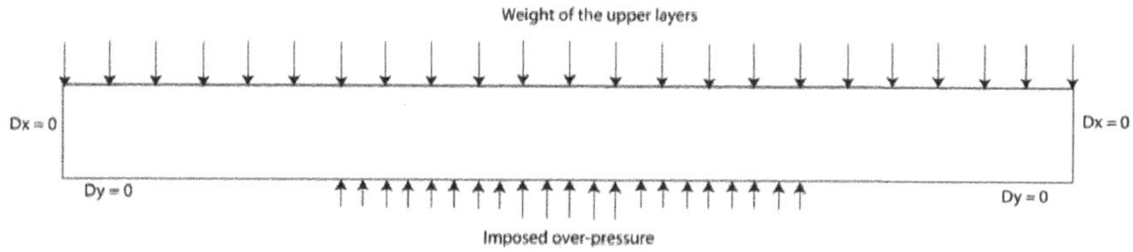

Figure 4.2. *Studied caprock and considered boundary conditions*

E (GPa)	v	K_{int} (m^2)	φ(%)	ρ (kg/m^3)	K_{IC} (MPa $m^{0.5}$)	m
16	0.28	10^{-17}	5	2,700	1.0	6

Table 4.1. *Principal hydromechanical parameters of the caprock*

When the maximum pressure reaches 2.1 times the initial pressure (p_0), tensile horizontal stresses appear on a part of the reservoir–caprock interface (Figure 4.3). First cracks are initiated when the prescribed pressure on the middle part of the model reaches $2.4p_0$. Figure 4.4 shows the corresponding pore pressure distribution in the caprock and the formed crack network. A zoom of the initiated crack network is presented in Figure 4.5. In this figure, the elements where the cracks are initiated are depicted in dark gray, propagation paths are black and finally the crack tips are in light gray. As it can be seen, all cracks are not initiated on the reservoir–caprock interface, where the horizontal stress is highest, due to the caprock heterogeneity.

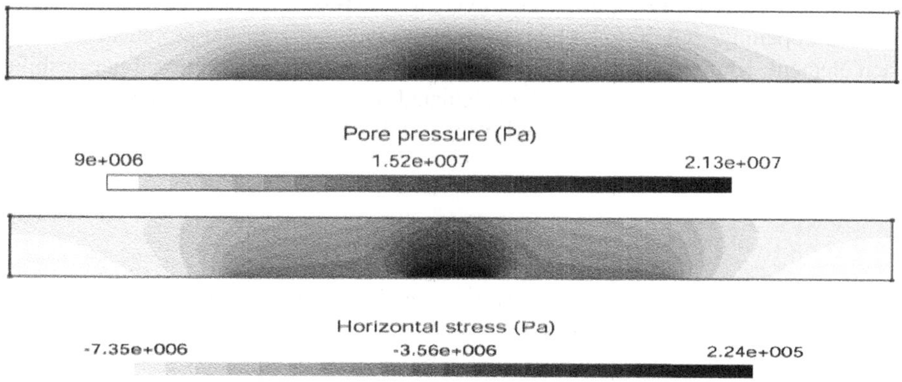

Figure 4.3. *Pore pressure distribution (top) and horizontal stresses (bottom) in the caprock expressed in Pa, when $p_i = 2.1p_0$*

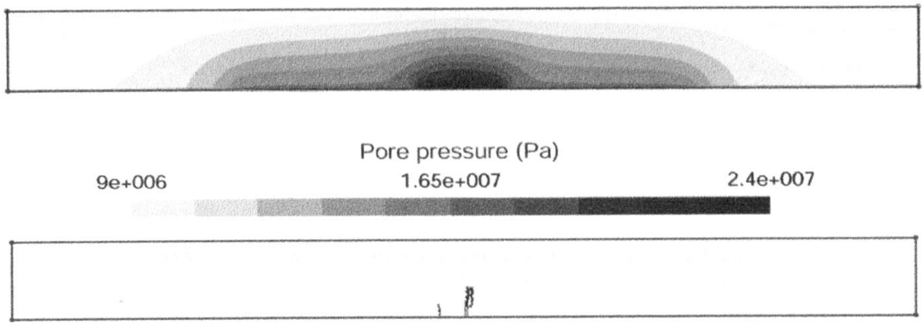

Figure 4.4. *Pore pressure distribution (top) in Pa and initiated crack network (bottom) in the caprock, when* $p_i = 2.4p_0$

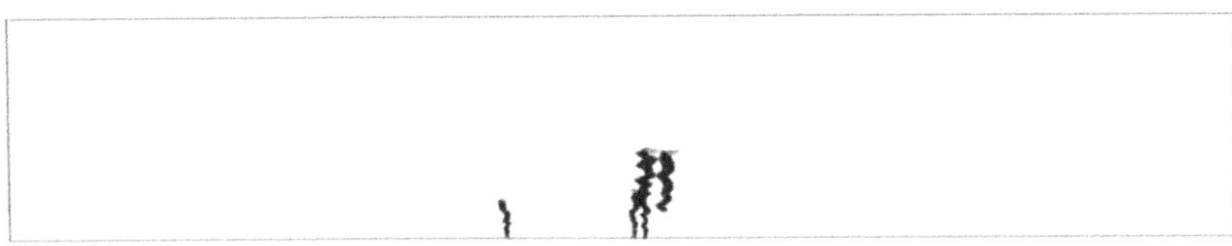

Figure 4.5. *Zoom on the formed crack network; dark gray elements show the initiation points, black elements propagation line and light gray elements crack tips*

The present simulations are stopped when the first crack network is formed, as the main goal is to estimate the sustainable injection pressure for a given site. No damage–permeability relationship is used. However, the non-local model correctly calculates the crack openings [GUY 12]. Only one macro-crack initiates and propagates through each element. Knowing the crack opening in each element, an additional permeability due to the presence of a crack can be calculated for each damaged element by using a cubic law as proposed by Olivella and Alonso [OLI 08]

$$K_d \cong K_{\text{int}} + \frac{b_f^3}{12a_f} \qquad [4.15]$$

where K_d is the equivalent permeability of a damaged element, K_{int} the reference intrinsic permeability of the rock matrix, b_f the opening of the fracture and a_f the width associated with each fracture, that is the element width. It is worth noting that taking into account the mechanical effects on the flow simulation and the use of a damage–permeability relationship can influence the results in terms of sustainable pressure and the pattern of crack network.

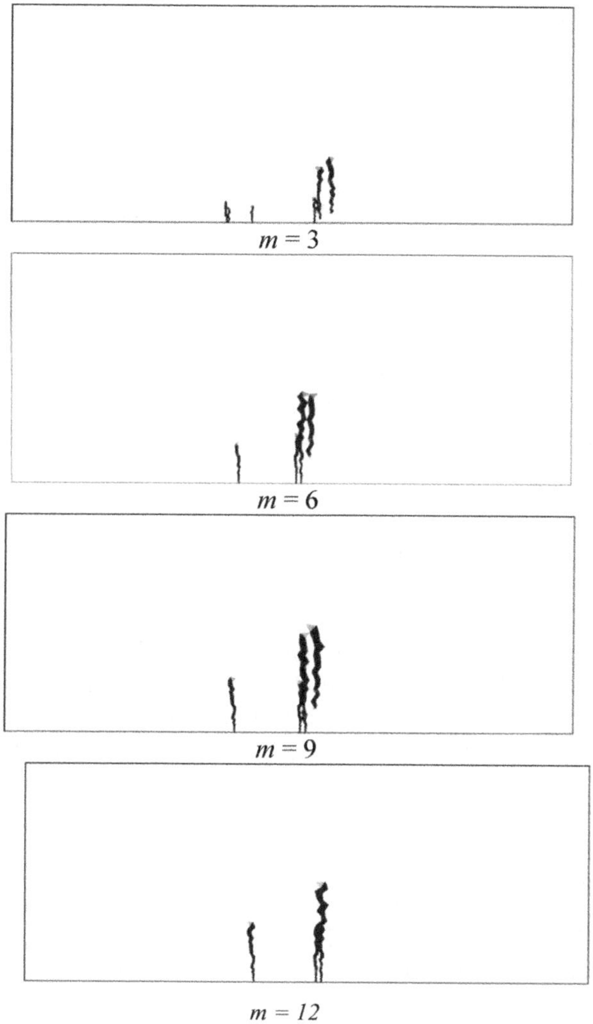

Figure 4.6. *Patterns of induced crack network for different values of m*

Weibull modulus m	Sustainable overpressure
3	$1.33p_0$
6	$1.4p_0$
9	$1.47p_0$
12	$1.47p_0$

Table 4.2. *Sustainable overpressure* $(p_i - p_0)$ *for different values of the Weibull modulus m*

4.4.1. *Effect of Weibull modulus*

The Weibull modulus, m, represents the heterogeneity of the studied material. A higher value of m corresponds to a less-heterogeneous material. The effect of m on the fracturing pressure of the studied caprock is investigated through a set of parametric calculations. Four typical values for m are considered, namely, $m = 3, 6, 9$ and 12. Figure 4.6 shows different crack networks obtained for four different values of m.

The results show that cracks can be initiated in a larger zone for more heterogeneous caprocks (i.e. small values of m). Furthermore, it is evident that the sustainable overpressure decreases slightly for more heterogeneous rocks as the first cracks are initiated earlier. In this particular case, the heterogeneity of the effective stress field attenuates the effects of material heterogeneity. It is worth noting that the parametric study is performed for different values of m while σ_0 and λ_0 remain constant.

4.5. Conclusions and perspectives

A non-local probabilistic model is used to simulate initiation and propagation of crack (fracture) networks in heterogeneous rocks. A stress regularization scheme is used in this framework to obtain mesh objective results. A natural transition from a damage mechanics-based model for crack initiation to a fracture mechanics-based model for crack propagation is thus performed. The model accounts for the interaction between propagating cracks and initiation sites.

The damage model is integrated in a hydromechanical sequential coupling scheme. A hydromechanical sequential coupling scheme is proposed and the pore pressure increases due to gas injection obtained from a hydraulic model that acts as a "load" for the mechanical model. The proposed scheme enables the effective stresses to be calculated directly. Crack initiation is checked at each time step considering regularized effective stresses. A set of parametric calculations is performed to illustrate the effects of rock heterogeneity on the sustainable overpressure and possible fracturing patterns. The results show that the sustainable overpressure decreases slightly for more heterogeneous caprocks.

The main goal of the presented study is to provide a methodology for evaluating the sustainable overpressure for a given site. High values of prescribed overpressure must be seen in this way. No specific damage–permeability coupling is taken into account herein. However, as the mechanical model calculates correctly the fracture openings, such a coupling can be considered in a straightforward manner by computing an additional permeability for damaged elements as a function of the opening of fractures.

4.6. Acknowledgments

Parts of the work presented here were supported by BRGM through an "Institut Carnot" research grant. Other parts were performed through EDF–BRGM research collaboration "ENDOSTON". The authors acknowledge the fruitful help of Dr. Sébastien Meunier from EDF R&D on numerical aspects of hydromechanical coupling.

4.7. Bibliography

[BAE 81] BAECHER G.B., EINSTEIN H.H., "Size effect in rock testing", *Geophysical Research Letters*, vol. 8, pp. 671–674, 1981.

[BIE 75] BIENIOWSKI Z.T., VAN HEERDEN W.L., "Significance of in-situ tests on large rock specimens", *International Journal of Rock Mechanics and Mining Sciences*, vol. 12, 1975, pp. 101–113.

[BIO 41] BIOT M.A., "General theory of three-dimensional consolidation", *Journal of Applied Physics*, vol. 12, pp. 155–164, 1941.

[BOR 95] BORJA R.I., ALARCON E., "A mathematical framework for finite strain elastoplastic consolidation. Part 1: balance laws, variational formulation, and linearization", *Computer Methods in Applied Mechanics and Engineering*, vol. 122, pp. 145–171, 1995.

[BOU 09] BOUC O., AUDIGANE P., BELLENFANT G., FABRIOL H., GASTINE M., ROHMER J., SEYEDI D., "Determining safety criteria for CO_2 geological storage", *Energy Procedia*, vol. 1, pp. 2439–2446, 2009.

[DEA 06] DEAN R.H., GAI S., STONE C.M., MINKOFF S.E., "A comparison of techniques for coupling porous flow and geomechanics", *SPE Journal*, vol. 11, pp. 132–140, 2006.

[EDF 08] Code_Aster, Code d'Analyses des Structures et Thermomécanique pour Etudes et Recherches, http://www.code-aster.org, 2008, EDF R&D.

[GAU 08] GAUS I., AUDIGANE P., ANDRE L., LION J., JACQUMET N., DURST P., CZERNICHOWSKI-LAURIOL I., AZAROUAL M., "Geochemical and solute transport modelling for CO_2 storage, what to expect from it?", *International Journal of Greenhouse Gas Control*, vol. 2, pp. 605–625, 2008.

[GUY 10] GUY N., SEYEDI D.M., HILD F., "Hydro-mechanical modelling of geological CO_2 storage and the study of possible caprock fracture mechanisms", *Georisk*, vol. 4, pp. 110–117, 2010.

[GUY 12] GUY N., SEYEDI D.M., HILD F., "A nonlocal probabilistic model for crack initiation and propagation in heterogeneous materials", *International Journal for Numerical Methods in Engineering*, vol. 90, pp. 1053–1072, 2012.

[KAN 85] KANNINEN M.F., POPELAR C.H., *Advanced Fracture Mechanics*, Oxford University Press, 1985.

[KIM 10] KIM J., Sequential methods for coupled geomechanics and multiphase flow, PhD thesis, Stanford University, 2010.

[OLI 08] OLIVELLA S., ALONSO E.E., "Gas flow through clay barriers", *Géotechnique*, vol. 58, pp. 157–176, 2008.

[PAR 83] PARK K. C., "Stabilization of partitioned solution procedure for pore fluid-soil interaction analysis", *International Journal for Numerical Methods in Engineering*, vol. 19, pp. 1669–1673, 1983.

[PEE 98] PEERLING R.H.J., de BORST R., BREKELMANS W.A.M., GEERS M.G.D., "Gradient-enhanced damage modeling of concrete fracture", *Mechanics of Cohesive-Frictional Materials*, vol. 38, pp. 323–342, 1998.

[PEE 01] PEERLING R.H.J., GEERS M.G.D., de BORST R., BREKELMANS W.A.M., "A critical comparison of non-local and gradient-enhanced softening continua", *International Journal of Solids and Structures*, vol. 38, pp. 7723–7746, 2001.

[PRU 02] PRUESS K., GARCIA J., "Multiphase flux dynamics during CO_2 disposal into saline aquifers", *Environmental Geology*, vol. 48, pp. 282–295, 2002.

[PRU 99] PRUESS K., OLDENBURG C.M., MORIDIS G.J. Tough2 user's guide, version 2.0, Lawrence Berkeley National Laboratory Report LBNL-43134, 1999, Berkeley, CA, USA.

[RUT 08] RUTQVIST J., BIRKHOLZER J.T., TSANG C., "Coupled reservoir-geomechanical analysis of the potential for tensile and shear failure associated with CO_2 injection in multilayered reservoir-caprock systems", *International Journal of Rock Mechanics and Mining Sciences*, vol. 45, pp. 132–143, 2008.

[ROH 10] ROHMER J., SEYEDI D.M., "Coupled large scale hydromechanical modelling for caprock failure risk assessment of CO_2 storage in deep saline aquifers", *Oil and Gas Science and Technology – Rev. IFP*, vol. 65, pp. 503–517, 2010.

[RUT 02a] RUTQVIST J., WU Y.-S., TSANG C.-F., BODVARSSON G., "A modeling approach for analysis of coupled multiphase fluid flow, heat transfer, and deformation in fractured porous rock", *International Journal of Rock Mechanics and Mining Sciences*, vol. 39, 2002, pp. 429–442.

[RUT 02b] RUTQVIST J., TSANG C.,-F., "A study of caprock hydromechanical changes associated with CO2 injection into a brine aquifer", *Environmental Geology*, vol. 42, 2002, pp. 296–305.

[SCH 04] SCHREFLER B.A., "Multiphase flow in deforming porous media", *International Journal for Numerical Methods in Engineering*, vol. 60, pp. 27–50, 2004.

[SET 98] SETTARI A., MOURITS F., "A coupled reservoir and geomechanical simulation system", *SPE Journal*, vol. 3, pp. 219–226, 1998.

[SET 01] SETTARI A., WALTERS D.A., "Advances in coupled geomechanical and reservoir modeling with applications to reservoir compaction", *SPE Journal*, vol. 6, pp. 334–342, 2001.

[SOL 09] SOLTANZADEH H., HAWKES C.D., "Assessing fault reactivation tendency within and surrounding porous reservoirs during fluid production or injection", *International Journal of Rock Mechanics and Mining Sciences*, vol. 46, pp. 1–7, 2009.

[VID 09] VIDAL-GILBERT S., NAUROY J.-F., BROOSE E., "3D geomechanical modelling for CO_2 geologic storage in the Dogger carbonates of the Paris Basin", *International Journal of Greenhouse Gas Control*, vol. 3, pp. 288–299, 2009.

Chapter 5

Caprock Breach: A Potential Threat to Secure Geologic Sequestration of CO_2

Geologic sequestration of CO_2 becomes feasible when stratigraphic configurations can promote the development of stable plumes of injected CO_2 within the storage horizon. In this regard, the caprock is an important component that will ensure the development of such plumes. The presence of a stable plume can enhance the activation of other trapping mechanisms, which require timescales substantially longer than the injection period. The injected fluids can, however, initiate interactions between the caprock and the surrounding geologic media, including both the storage formation and the overburden rocks. The interaction can lead to the initiation of damage and fracture that can pose a threat to geologic sequestration. This chapter presents recent advances in the modeling of the interaction between the caprock and the surrounding geologic media under the action of a disk-shaped plume of uniform pressure located at a finite depth from the caprock. This chapter also discusses the implications of a fracture generated during an injection scenario and the attendant steady leakage rates that are influenced by the permeabilities of the system.

5.1. Introduction

The deep geologic sequestration of CO_2 in supercritical form in geologic settings that are conducive to the development of stable injected plumes is regarded as an essential requirement for successful implementation of carbon capture and secure

Chapter written by A.P.S. SELVADURAI.

storage (CCSS). Stable plumes can enhance the activation of other trapping mechanisms, which require timescales substantially longer than the injection period. For example, if CO_2 injection lasts 50 years, primary trapping mechanisms such as adsorption, structural and stratigraphic trap filling, and hydrodynamic trapping are expected to commence immediately and last up to a million years; secondary trapping mechanisms such as residual CO_2 trapping can peak at around 10,000 years and continue thereafter; dissolution can peak around a thousand years and last up to a million years, while mineralization can continue beyond a million years. To effectively activate all trapping mechanisms, it is essential that the injected CO_2 remains in a storage formation in a hydrodynamically stable condition. There are a number of processes and conditions that can potentially result in the development of adverse conditions, which can compromise the secure storage potential of a geological formation (Figure 5.1).

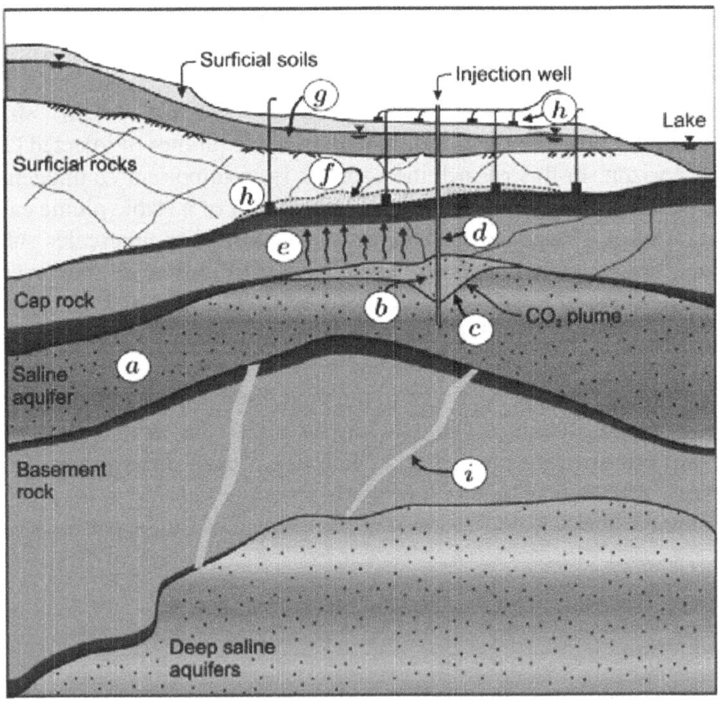

Figure 5.1. *Schematic view of $scCO_2$ injection into a saline aquifer*

Referring to Figure 5.1, the research themes relevant to assessing secure geologic storage could include a) geochemistry and geomechanics of a virgin saline storage formation; b) plume development, trapping, and pore fabric alterations during injection; c) stability of supercritical CO_2 ($scCO_2$)–saline pore fluid interfaces in porous media; d) geochemistry and geomechanics of caprock and interface seals; e) CO_2 diffusion and transport through caprock defects and interfaces in access boreholes; f) plume development in surficial rocks; g) CO_2 contamination of groundwater; h) caprock deformations and uplift; and i) movement of saline fluids from deep saline aquifers to the storage formation.

The literature on geologic sequestration of CO_2 published over the past four decades has placed a great deal of emphasis on the fluid dynamics aspects of the sequestration problem, with an emphasis on advective transport, buoyancy effects associated with density mismatch, and geochemical actions during transport. The literature covering these topics is vast and no attempt will be made to provide a compendium of relevant papers. The articles and volumes (e.g. [HOL 01, BAC 03, SEL 03a, SEL 06, SEL 12a, PRU 04, OEL 06, AND 07, WIL 07, LAL 08, LEW 09, MCP 09]) provide references to further studies. As is evident, the great majority of approaches to the geologic sequestration of CO_2 treat the problem purely from the context of fluid transport and geochemistry, and the geomechanical aspects of the interaction between the injected fluids and the storage rock and the confining caprock is gaining attention (e.g. [RUT 02, RUT 07, RUT 08, SEL 09, SEL 12b, KRA 12]). The assessment of leakage of stored fluids from defects created in caprock and from access boreholes and other distributed defects has been examined (e.g. [SEL 11, SEL 12a]). In this chapter, we present the results of recent studies that provide convenient estimates for examining the state of stress in a caprock layer that is embedded within a storage formation and a surficial geologic stratum and the results for fluid leakage from defects that can be created in an embedded caprock layer.

5.2. Caprock flexure during injection

We consider the specific problem of the interaction between a caprock layer (subscript "c") that is embedded in bonded contact between an overburden rock (subscript "o") and a storage formation (subscript "s") (Figure 5.2). As a first approximation, the mechanical behavior of all geological formations are modeled by isotropic elasticity. The overburden and storage layers are considered to be half-space regions and the caprock layer is modeled according to the Germain–Poisson–Kirchhoff thin plate theory (e.g. [TIM 59, SEL 79, SEL 00a]) although the analysis can be extended to include shear deformations of the plate (e.g. [REI 45, RAJ 85]). The elastic interaction between the storage formation, the caprock, and the overburden rock is induced by fluids that are injected under pressure into the storage formation. For convenience, the pressurization zone is assumed to be a disk-shaped region of finite radius and finite thickness and located at a finite distance from the caprock. The resulting elastic interaction problem is axisymmetric (Figure 5.2).

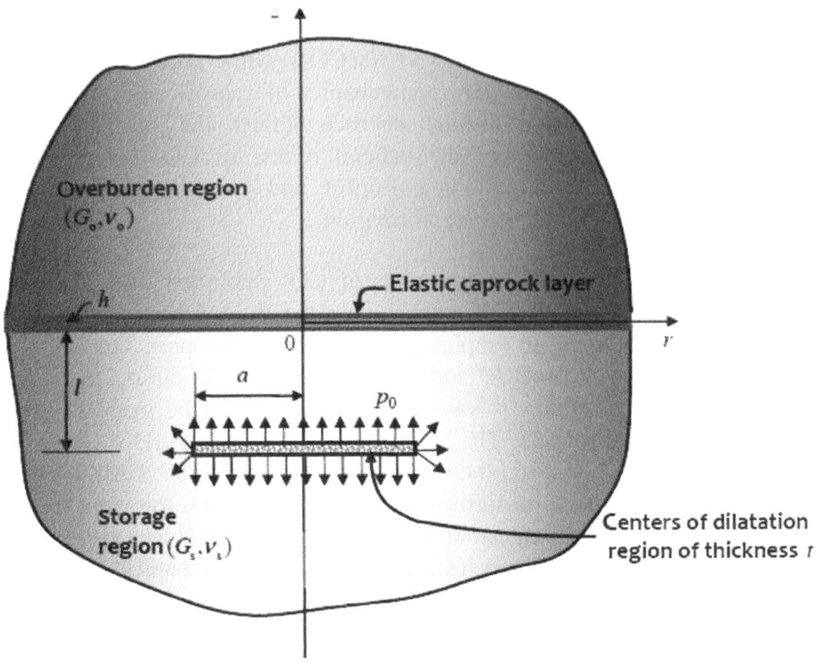

Figure 5.2. *Caprock–storage formation–overburden rock interaction due to an injected region of pressurized fluid*

For the mathematical analysis, the problem is examined by considering the flexural interaction between the caprock layer and the storage formation and the overburden layer, which is induced by the disk-shaped injection region. The differential equation governing the flexural deflection $w(r)$ of the caprock layer is given by [SEL 79]

$$D\tilde{\nabla}^2\tilde{\nabla}^2 w(r) + q^{(s)}(r) - q^{(o)}(r) = 0 \qquad [5.1]$$

where

$$\tilde{\nabla}^2 = \frac{d^2}{dr^2} + \frac{1}{r}\frac{d}{dr} \qquad [5.2]$$

In [5.1], $q^{(s)}(r)$ and $q^{(o)}(r)$ are, respectively, the contact stresses acting at the storage formation–caprock and the overburden–caprock interfaces. Also, $D(=G_c h^3/6(1-\nu_c))$ is the flexural rigidity of the caprock layer, where h is its thickness; G_C and v_C are, respectively, its linear elastic shear modulus and Poisson's ratio. We assume that there is no separation between the caprock layer and the storage and overburden regions, and the elastic plate model selected to represent the caprock layer imposes an inextensibility constraint at the interface regions thereby imposing the boundary conditions

$$w(r) = u_z^{(s)}(r,0) + u_z^{(s)p_0}(r,0) = u_z^{(o)}(r,0) \qquad [5.3]$$

$$u_r^{(s)}(r,0) = u_r^{(s)p_0}(r,0) = u_r^{(o)}(r,0) = 0 \qquad [5.4]$$

where $u_z^{(s)}$, $u_z^{(o)}$ and $u_z^{(s)p_0}$ are the axial displacements at the surfaces of the half-space regions due to the contact stresses and the pressurized zone and $u_r^{(s)}$, $u_r^{(o)}$ and $u_r^{(s)p_0}$ are the appropriate radial displacements. Using a Hankel transform development of the governing equations of elasticity (e.g. [SNE 51, SEL 00a]), it can be shown that

$$u_z^{(s)p_0}(r,0) = \frac{\alpha_s(1-2v_s)p_0 at}{2G_s(1-v_s)} \int_0^\infty e^{-\xi l} J_1(\xi a) J_0(\xi r)\, d\xi \qquad [5.5]$$

where $J_0(x)$ and $J_1(x)$ are, respectively, the zeroth-order and first-order Bessel functions of the first kind and α_s is the Biot coefficient of the storage formation defined by

$$\alpha_s = 1 - \frac{\tilde{K}}{\tilde{K}_g} \qquad [5.6]$$

where $\tilde{K}$ is the compressibility of the skeleton of the storage medium and $\tilde{K}_g$ is the compressibility of the grain material. Similarly, it can be shown that the Hankel transform equivalent to the deflection of the plate

$$\overline{w}(\xi) = \int_0^\infty \xi\, w(r) J_0(\xi r)\, d\xi \qquad [5.7]$$

is related to the Hankel transforms of the axial displacements in the storage and overburden regions $\overline{u}_z^{(s)}(\xi)$, $\overline{u}_z^{(p_0)}(\xi)$, and $\overline{u}_z^{(o)}(\xi)$ through the relations

$$\overline{w}(\xi) = \overline{u}_z^{(s)}(\xi) + \overline{u}_z^{(s)p_0}(\xi) = \frac{(3-4v_s)}{4G_s(1-v_s)\xi}[\overline{q}^{(s)}(\xi) - \overline{S}(\xi)] \qquad [5.8]$$

$$\bar{w}(\xi) = \frac{(3-4\nu_o)}{4G_o(1-\nu_o)\xi}\bar{q}^{(o)}(\xi) \qquad [5.9]$$

where

$$\bar{S}(\xi) = \frac{p_0 a t \alpha_s (1-2\nu_s)}{(3-4\nu_s)} e^{-\xi l} J_1(\xi a) \qquad [5.10]$$

Considering the transformed equivalent of [5.1], [5.8] and [5.9], it can be shown that the integral representation for the deflected shape of the caprock due to the pressurized injection region of radius a and thickness t that is located at a depth l from the caprock can be evaluated in the form

$$w(r) = \frac{\Omega a t}{h} \int_0^\infty \frac{e^{-(\lambda l/h)}}{[1+\Phi\lambda^3]} J_1(\lambda a/h) J_0(\lambda r/h)\, d\lambda \qquad [5.11]$$

where

$$\Omega = \frac{\alpha_s (1-\nu_s)(1-2\nu_s)(3-4\nu_s) p_0}{2[G_s(1-\nu_s)(3-4\nu_o) + G_o(1-\nu_o)(3-4\nu_s)]} \qquad [5.12]$$

$$\Phi = \frac{G_c(3-4\nu_s)(3-4\nu_o)}{24(1-\nu_c)[G_s(1-\nu_s)(3-4\nu_o) + G_o(1-\nu_o)(3-4\nu_s)]} \qquad [5.13]$$

A more generalized result that takes into account injection within both the storage formation and overburden rocks and possible influences of slippage at the interfaces will be given elsewhere [SEL 12a]. The analytical solution for the deflection of the embedded caprock and the adjacent geologic media [5.11] can be used in the expressions that follow to determine the flexural moments and shear forces in the caprock layer e.g.

$$M_{rr}(r) = -D\left[\frac{d^2 w(r)}{dr^2} + \frac{\nu_c}{r}\frac{dw(r)}{dr}\right] \qquad [5.14]$$

$$M_{\theta\theta}(r) = -D\left[\frac{1}{r}\frac{dw(r)}{dr} + \nu_c \frac{d^2 w(r)}{dr^2}\right] \qquad [5.15]$$

$$V_r(r) \quad = -D\frac{d}{dr}\left[\frac{d^2 w(r)}{dr^2} + \frac{1}{r}\frac{dw(r)}{dr}\right] \qquad [5.16]$$

5.2.1. *Numerical results for the caprock–geologic media interaction*

The result [5.11] contains two non-dimensional parameters Ω and Φ, which can be evaluated by assigning typical relative values for the elasticity, geometrical and loading parameters in the problem. The quantity $(\Omega h / at)$ can be used to non-dimensionalize $w(r)$ and the influence of Φ, and the geometrical parameters l / h, a/h, and r/h can be assigned typical values. For example, consider an injection zone which contains a caprock layer of thickness $h = 50$ m. The storage region is pressurized over a radius $a = 500$ m with thickness $t = 50$ m located at a depth of $l = 250$ m from the boundary of the caprock. The value of l/h can be varied to examine the influence of the position of the injection zone on the caprock displacement. Also, in the numerical results presented, we set $\Phi = 2$, which corresponds to $(G_c / (G_s + G_o) \approx 13.5$ and $\nu_C = \nu_S = \nu_O = 1/4$. The result [5.11] is evaluated using MATHEMATICA™ and the upper limit of the integral is selected to ensure accuracy of the infinite integral. Figures 5.3 to 5.5 illustrate the deflected shapes for the caprock for $a / h = 10$ and for $\Phi \in (0.1, 100)$.

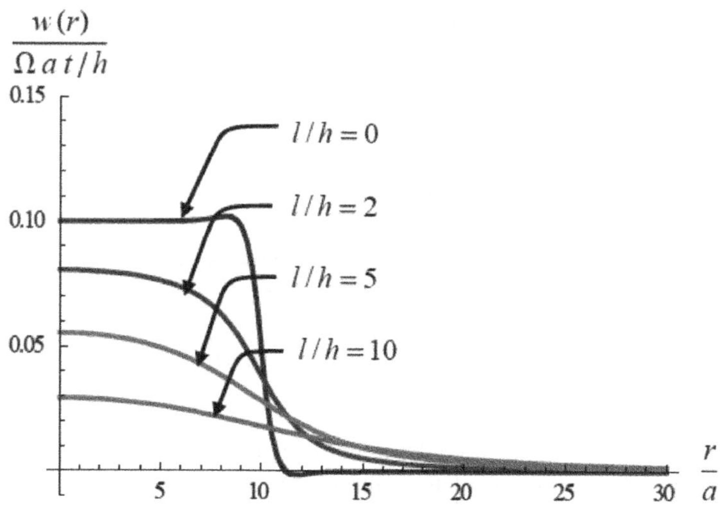

Figure 5.3. *Deflection of the embedded caprock layer due to a disk-shaped pressurized region located in the storage rock ($a / h = 10$; $\Phi = 0.10$)*

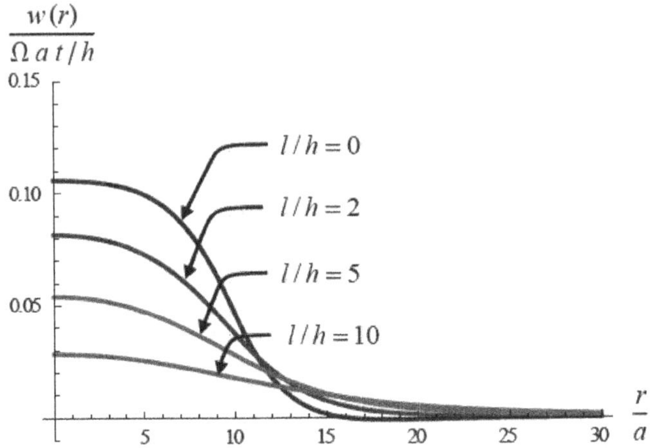

Figure 5.4. *Deflection of the embedded caprock layer due to a disk-shaped pressurized region located in the storage rock* $(a/h = 10;\ \Phi = 10)$

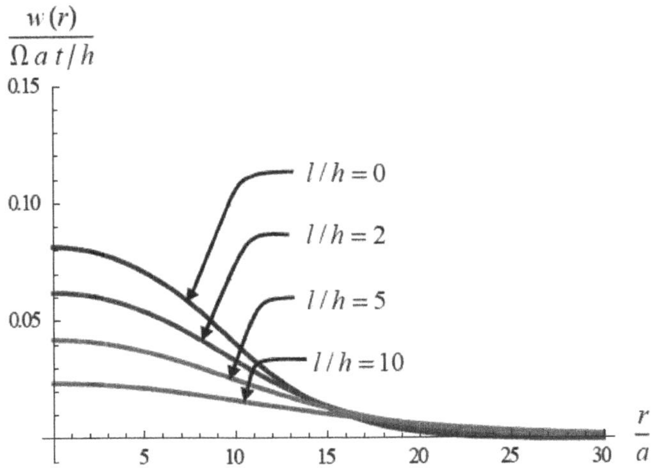

Figure 5.5. *Deflection of the embedded caprock layer due to a disk-shaped pressurized region located in the storage rock* $(a/h = 10;\ \Phi = 100)$

The numerical results illustrate the importance of the location of the pressurized region on the deflected shape of the caprock. As the thickness of the overburden region becomes small, the caprock deflections can manifest as surface deflections of the injection site. Pinsar Satellite observations at the In Salah CO_2 sequestration site in Algeria [VAS 08] indicate the type of non-uniform ground heave that is associated with the injection of CO_2 (Figure 5.6).

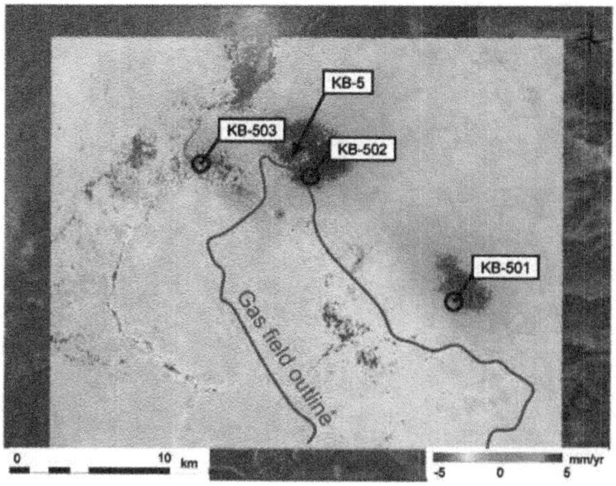

Figure 5.6. *Satellite observations of ground heave at the In Salah CO_2 sequestration site in Algeria [VAS 08]*

The integral expression [5.11] for the deflection of the caprock layer can be used in the result [5.14] to evaluate the radial flexural moment in the caprock layer. Figures 5.7 to 5.9 illustrate the variation in the radial flexural moment M_{rr} with the radial coordinate, for $a / h = 10$ and for $\Phi \in (0.1, 100)$. It is clear that, as the relative rigidity of the caprock increases, the location at which the maximum flexural moment occurs shifts from the edges of the pressurized region (Figure 5.7) toward its center (Figure 5.9). This process can influence the location and type of defect generation in the caprock layer.

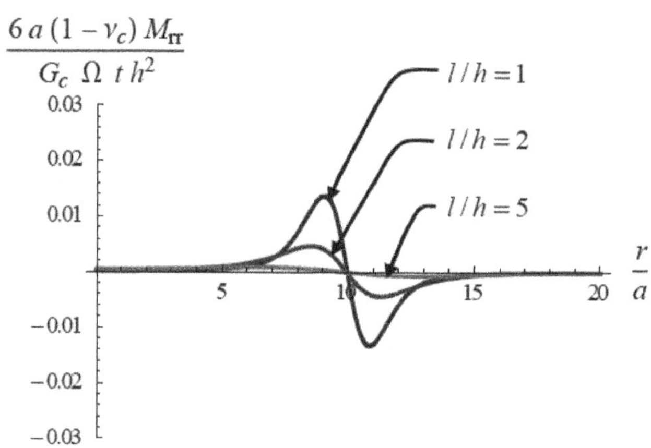

Figure 5.7. *Variation of radial flexural moment in the caprock layer due to internal pressurization of the storage region ($a / h = 10$; $\Phi = 0.10$)*

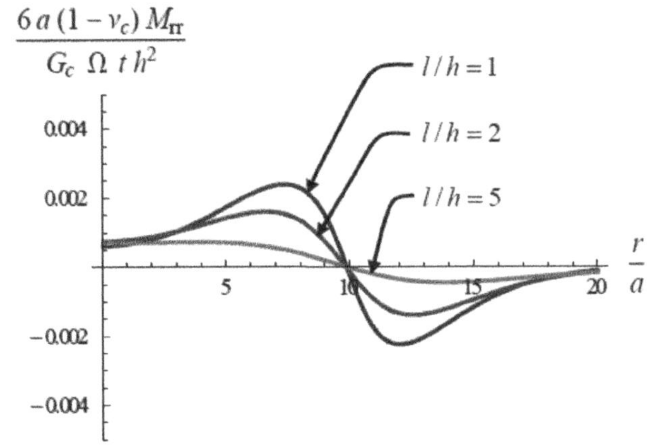

Figure 5.8. *Variation of radial flexural moment in the caprock layer due to internal pressurization of the storage region* $(a/h=10;\ \Phi=10)$

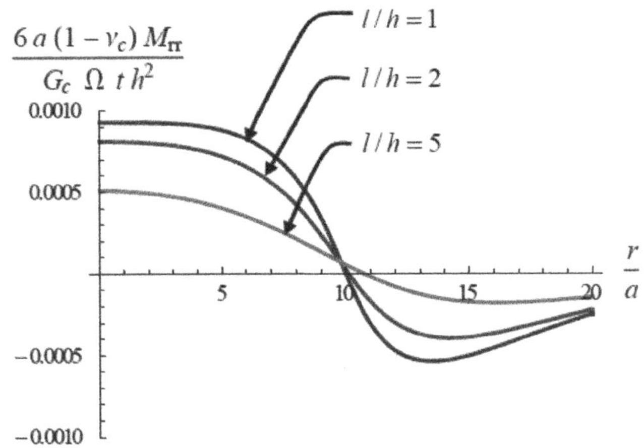

Figure 5.9. *Variation of radial flexural moment in the caprock layer due to internal pressurization of the storage region* $(a/h=10;\ \Phi=100)$

The radial moment can be used to assess the tensile flexural stresses that can be developed on the boundary faces of the caprock layer. The nature of the internal pressurization is such that the caprock will experience compressive and tensile stresses at these boundaries. The *in situ* stress state consistent with the depth of embedment of the caprock layer combined with the flexural stress state induced by the injected fluids needs to be assessed to establish the permissible injection pressures that will not cause failure or crack initiation in the caprock according to a prescribed failure criterion [DAV 02].

5.3. Fluid leakage from a fracture in the caprock

Some control on the initiation of fractures and damage zones in a caprock layer can be exercised through adequate control of the injection pressures and knowledge of the geometry of the injection zone. The activation of defects that can go undetected during a site investigation exercise that uses either seismic investigations or borehole injection techniques presents a greater threat to storage security since leakages into the groundwater regime in the overburden could go undetected leading to possible groundwater contamination. It is also unrealistic to expect that sequestration sites free from any defects could ever be selected for geologic sequestration of CO_2 in supercritical form. It is therefore prudent to anticipate some activation of fractures (i.e. either creation of new fractures, opening of dormant fractures, or extension of existing fractures) and to establish leakage rates through cracks, particularly when sequestration activities are in progress, so that their risk to storage security can be estimated. We examine the problem of the fluid leakage problem for a crack located in the caprock. Attention is restricted to the specific problem where leakage takes place through a vertically aligned caprock fracture with an elliptical cross-section, connecting a storage region of permeability K_s and an overburden region of permeability K_o. The storage rock and the overburden rock are maintained at constant reduced far-field Bernoulli potentials Φ_s and Φ_o, respectively, with $\Phi_s > \Phi_o$ (Figure 5.10). The flow of fluids within the porous media is described by Darcy's law. Although the viscosity properties and other flow characteristics of the injected $scCO_2$ are different from the resident fluids in a storage formation, we assume that there is no mismatch in the flow characteristics of the resident and injected fluids. Furthermore, attention is restricted to steady flow in the system, although the methodologies can be extended to include transient effects and thermal phenomena.

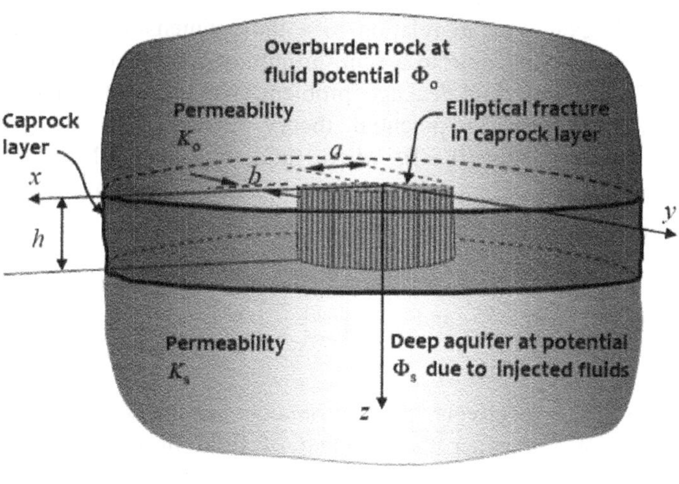

Figure 5.10. *Caprock fracture in a geologic storage setting*

For Darcy flow in an isotropic and homogeneous porous medium, the average flow velocity $\mathbf{v}(\mathbf{x})$ in the medium satisfies the relationship

$$\mathbf{v}(\mathbf{x}) = -\frac{K\gamma_w}{\mu}\nabla\Phi(\mathbf{x}) \qquad [5.17]$$

where K is the scalar permeability, γ_w is the unit weight of the fluid, μ is the dynamic viscosity, ∇ is the gradient operator, $\Phi(\mathbf{x})$ is the scalar Bernoulli potential, and $\mathbf{x}$ is the spatial variable. In general, the permeability is a *second-order tensor*, which gives rise to directional properties with spatial variability. Such multi-directional and multi-scale extensions are possible (e.g. [SEL 03b, SEL 04, SEL 07, SEL 10, SEL 11]), but are excluded from the ensuing analysis. For incompressibility of the permeating fluid and non-deformability of the porous skeleton, the equation of mass conservation takes the form

$$\nabla \cdot \mathbf{v}(\mathbf{x}) = 0 \qquad [5.18]$$

The partial differential equation governing fluid flow in the porous medium obtained by combining [5.17] and [5.18] gives rise to Laplace's equation

$$\nabla^2\Phi(\mathbf{x}) = 0 \qquad [5.19]$$

A solution of [5.19], which is subject to the relevant Dirichlet and Neumann boundary conditions prescribed on complementary subsets of the relevant boundaries of a fluid flow domain, constitutes a unique solution [SEL 00b]. Since the fluid flow is examined in relation to a three-dimensional configuration, the regularity conditions applicable to a semi-infinite domain should also be satisfied. For the solution of the fluid leakage problem, we focus attention on a mixed boundary value problem in potential theory related to a half-space region $(x \in (-\infty,\infty); y \in (-\infty,\infty); z \in (0,\infty))$, where fluid flow takes place in an elliptical aperture of an otherwise sealed surface. We define the region within (S_i) and exterior (S_e) to the plane elliptical region as follows:

$$\left(\frac{x}{a}\right)^2 + \left(\frac{y}{b}\right)^2 \leq 1; \;\; S_i \quad \text{and} \quad \left(\frac{x}{a}\right)^2 + \left(\frac{y}{b}\right)^2 > 1; \;\; S_e \qquad [5.20]$$

where a and b are, respectively, the semi-major and the semi-minor axes of the elliptical region. We consider the problem of fluid flow through the elliptical aperture contained at the boundary of the impervious caprock that is in contact with the porous half-space region. For the analysis of the fluid flow problem, we consider the mixed boundary conditions (Figure 5.11):

$$(\Phi)_{z=0} = \varphi_0 = \text{const.}; \quad (x,y) \in S_i \qquad [5.21]$$

$$\left(\frac{\partial \Phi}{\partial z}\right)_{z=0} = 0; \quad (x,y) \in S_e \qquad [5.22]$$

The method of solution for the mixed boundary value problem, which uses the ellipsoidal harmonic function technique, is described elsewhere (e.g. [LAM 27, MOR 53, GRE 50, KAS 68, SEL 82, SEL 10, SEL 12b]) and will not be repeated here.

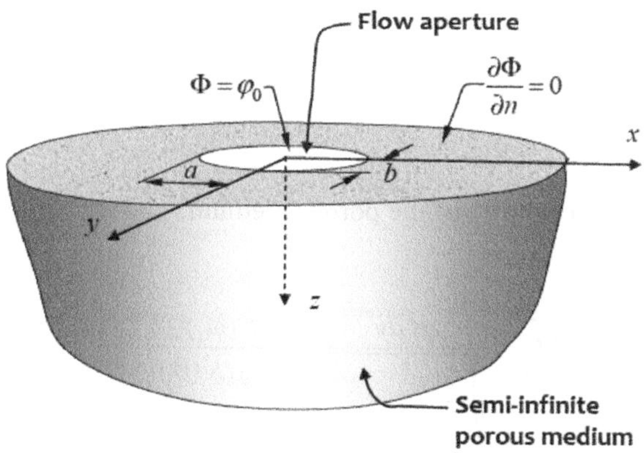

Figure 5.11. *Boundary conditions at the interface between the porous medium and an impervious rock layer containing an elliptical aperture*

It is sufficient to note that the integral expression for the potential that satisfies the mixed boundary conditions described by [5.21] and [5.22] can be written in the form

$$\Phi(x,y,z) = \frac{a\,\varphi_0}{K(\sigma)} \int_{\xi}^{\infty} \frac{ds}{\sqrt{s(a^2+s)(b^2+s)}} \qquad [5.23]$$

where $K(\sigma)$ is the complete elliptic integral of the first kind, defined by

$$K(\sigma) = \int_0^{\pi/2} \frac{d\varsigma}{\sqrt{1-\sigma^2 \sin^2 \varsigma}}; \qquad \sigma = \left(\frac{a^2-b^2}{a^2}\right)^{1/2} \qquad [5.24]$$

$$\xi = a^2[\mathrm{sn}^{-2}(u,\sigma)-1] \qquad [5.25]$$

and $\mathrm{sn}(u,\sigma)$ represents the Jacobian elliptic function defined through the integral

$$\int_0^{\mathrm{sn}(u,\sigma)} \frac{dt}{\sqrt{(1-t^2)(1-\sigma^2 t^2)}} = (u,\sigma) \qquad [5.26]$$

Tabulated values for all Jacobian elliptic functions are given in the literature (e.g. [ABR 64]).

The fluid velocity at the elliptical aperture is given by

$$v_z(x,y,0) = -\frac{K^* \gamma_w}{\mu}\left(\frac{\partial \Phi}{\partial z}\right)_{z=0} = \frac{K^* \gamma_w \varphi_0}{\mu b\, K(\sigma)} \frac{1}{\sqrt{1-\frac{x^2}{a^2}-\frac{y^2}{b^2}}} \qquad [5.27]$$

where K^* is the permeability of the porous medium. The flow rate into the elliptical aperture is given by

$$Q = \frac{K^* \gamma_w \varphi_0}{\mu b\, K(\sigma)} \iint_{S_i} \frac{dx\, dy}{\sqrt{1-\frac{x^2}{a^2}-\frac{y^2}{b^2}}} = \frac{2\pi a \varphi_0 \gamma_w K^*}{\mu K(\sigma)} \qquad [5.28]$$

It should be noted that although the velocity at the boundary of the elliptical entry region is infinite, the volume flow rate to the entry region is finite.

Consider the problem of flow from the storage region toward the elliptical aperture region, corresponding to one end of the fracture, located at the impermeable rock interface. The far-field reduced Bernoulli potential is denoted by Φ_S and the corresponding value at the elliptical aperture is denoted by φ_S, with $\Phi_S > \varphi_S$. The flow velocity at the elliptical aperture is now given by [5.27], with the value of φ_0 being simply replaced by $(\Phi_S - \varphi_0)$. These results are sufficient to examine the problem of fluid flow through the fracture in a caprock layer embedded between two semi-infinite porous regions. Identical results can be developed for the flow within the overburden rock that is maintained at a far-field potential Φ_O and the potential at the elliptical region through which flow takes place is denoted by φ_O. The flow problem can be solved by making use of the assumption of continuity of flow between the storage formation and the overburden geologic material which takes place through the elliptical crack. The permeability of the elliptical crack can be estimated by considering the problem of viscous Poiseulle flow in a prismatic cylinder with an elliptical cross-section (e.g. [LAN 64]). Considering the problem where the plane ends of the prismatic cylinder are subjected to constant potentials over the elliptical areas, it can be shown that the permeability of the elliptical prismatic fracture is given by

$$K_F = \frac{a^2 b^2}{4(a^2 + b^2)} \quad [5.29]$$

where $2a$ is the major axis of the ellipse and $2b$ is the minor axis. The solution of the steady flow problem can be obtained by making use of the continuity equation $Q_s = Q_o = Q_F = Q$, which gives

$$Q = \left(\frac{2\pi ab\,\gamma_w K_F K_s K_o}{\mu\,[bK_F(K_o + K_s)K(\sigma) + 2hK_s K_o]} \right) (\Phi_s - \Phi_o) \quad [5.30]$$

5.3.1. *Numerical results for fluid leakage from a fracture in the caprock*

The result obtained from [5.30] can be written in the following non-dimensional form:

$$\bar{Q} = \frac{Q\mu}{2\pi(\Phi_s - \Phi_o)\gamma_w a K_s} = \frac{\Gamma\Delta}{\left(\Gamma\Delta[1+\Omega]K(\sigma) + 2\left(\frac{h}{a}\right) \right)} \quad [5.31]$$

where

$$\mathrm{M} = \frac{K_s}{K_o}; \quad \Delta = \frac{K_F}{K_s}; \quad \Gamma = \frac{b}{a}; \quad \sigma = \sqrt{1 - \Gamma^2} \quad [5.32]$$

Since the analytical result for the steady leakage rate from the crack can be evaluated in a particularly simple form, it can be used to develop a variety of solutions involving leakage from fracture swarms, damaged regions, etc. An extensive study of this problem was recently presented [SEL 12a]. Figure 5.12 presents typical results for the variation of $\bar{Q}$ with the permeability mismatch K_S / K_O between the storage and overburden rocks, the permeability mismatch between the fracture and the storage region K_F/K_S, and the relative geometry of the crack h/a. Figure 5.12 also illustrates (open circles) the flow rates through an elliptical fracture determined from a computational approach using the COMSOL™ code.

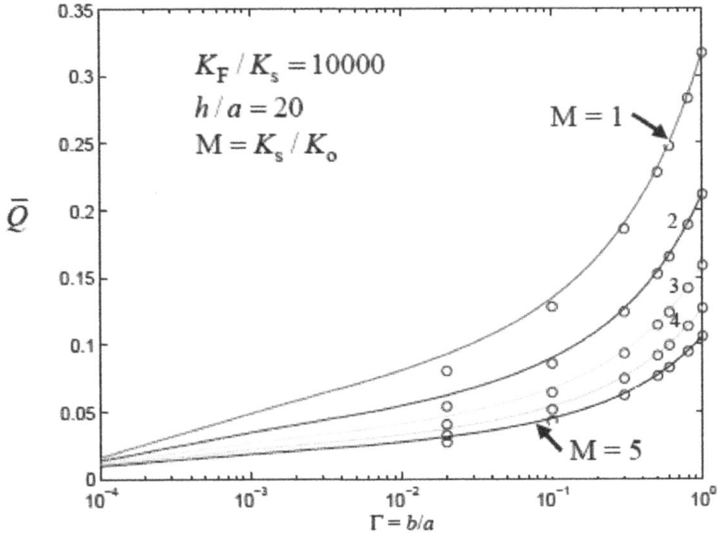

Figure 5.12. *Steady leakage rates from storage formation to overburden rocks through a vertically aligned elliptical crack*

5.4. Concluding remarks

Issues related to the longevity and security of CO_2 stored in geological formations are now being discussed as an essential part of the overall strategy to ensure that the stored fluids will not pose a threat to groundwater resources. The action of injected CO_2 in supercritical form needs to be treated as a geomechanical problem where the thermo-hydro-mechanical-chemical (THMC) processes that can occur both during and post injection can be better understood. The treatment of the storage problem as a groundwater hydrology problem raises concerns, particularly in situations where the integrity of the primary geological barrier can be compromised by mechanical actions. The possibility of inducing fracture and crack initiation in the caprock needs to be examined using sophisticated THM and THMC models that have been developed for other endeavors such as the geologic sequestration of hazardous materials. The most viable approach to examining caprock integrity will require the use of computational schemes that incorporate THMC effects. The scope of such analyses also requires the availability of well-documented geomaterial characterizations of the storage, caprock, and overburden horizons. In preliminary assessments of the feasibility of potential sites, such information is scarce. Therein exists the justification for developing elementary solutions to geomechanical interactions that can materialize during pressurized injection into a storage setting. This chapter summarizes recent advances in this area, particularly as they relate to flexure of an embedded caprock layer during CO_2 injection. The potential for fracture initiation in the caprock can be examined through the types of solutions presented in this chapter. This chapter also proposes techniques that can be used to establish the leakage rates that can occur in the event that a fracture is generated in the caprock or a pre-existing defect is activated. These elementary approaches can not only be used in preliminary calculations of plausible effects in storage horizons but can also be used as benchmarking solutions for validating computational approaches.

5.5. Acknowledgment

The work presented in this chapter shows the ongoing research in environmental geomechanics being conducted at the Department of Civil Engineering and Applied Mechanics, McGill University, and supported through research grants awarded by NSERC. The author is also grateful to the Max Planck Gesellschaft, Bonn, Germany, which provided research support in the form of the *Max Planck Research Prize in the Engineering Sciences*, 2003–2009.

5.6. Bibliography

[ABR 64] ABRAMOWITZ M., STEGUN I.A., *Handbook of Mathematical Functions with Formulas, Graphs and Mathematical Functions*, Dover, New York, 1964.

[AND 07] ANDRÉ L., AUDIGANE P., AZAROUAL M., MENJOZ A., "Numerical modeling of the fluid-rock chemical interaction at the supercritical CO_2-liquid interface during CO_2 injection into a carbonate reservoir, the Dogger aquifer (Paris Basin, France)", *Energy Conservation and Management*, vol. 48, pp. 1782–1797, 2007.

[BAC 03] BACHU S., ADAMS J.J., "Sequestration of CO_2 in geologic media in response to climate change: capacity of deep saline aquifers to sequester CO2 in solution", *Energy Conservation and Management*, vol. 44, pp. 3151–3175, 2003.

[DAV 02] DAVIS R.O., SELVADURAI A.P.S., *Plasticity and Geomechanics*, Cambridge University Press, Cambridge, 2002.

[GRE 50] GREEN A.E., SNEDDON I.N., "The distribution of stress in the neighbourhood of a flat elliptical crack in an elastic solid", *Proceedings of the Cambridge Philosophical Society*, vol. 46, pp. 159–163, 1950.

[HOL 01] HOLLOWAY S., "Storage of fossil fuel-derived carbon dioxide beneath the surface of the earth", *Annual Review of Energy and the Environment*, vol. 26, pp. 145–166, 2001.

[KAS 68] KASSIR M.K., SIH G.C., "Some three-dimensional inclusion problems in elasticity", *International Journal of Solids and Structures*, vol. 4, pp. 225–241, 1968.

[KRA 12] KRAMER D., "Scientists poke holes in carbon dioxide sequestration", *Physics Today*, vol. 65, no. 8, pp. 22–24, 2012.

[LAL 08] LAL R., "Carbon sequestration", *Philosophical Transactions of the Royal Society, B*, vol. 363, pp. 815–830, 2008.

[LAM 27] LAMB H., *Hydrodynamics*, 6th ed., Cambridge University Press, Cambridge, 1927.

[LAN 64] LANGLOIS W., *Slow Viscous Flow*, Macmillan, New York, 1964.

[LEW 09] LEWICKI J.L., BIRKHOLZER J., TSANG C.-F., "Natural and industrial analogues for leakage of CO_2 from storage reservoirs: identification of features, events and processes and lessons learned", *Environmental Geology*, vol. 52, pp. 457–467, 2009.

[MCP 09] MCPHERSON B.J.O.L., SUNDQUIST E.T. (eds), *Carbon Sequestration and its Role in the Global Carbon Cycle*, AGU Monograph Series, American Geophysical Union, Washington, DC, 2009.

[MOR 53] MORSE P.M., FESHBACH H., *Methods of Theoretical Physics, 1 and 2*, McGraw-Hill, New York, 1953.

[OEL 06] OELKERS E.H., COLE D.R., "Carbon dioxide sequestration. A solution to a global problem", *Elements*, vol. 4, pp. 05–310, 2006.

[PRU 04] PRUESS K., GARCIA J., KOVSCEK T., OLDENBURG C., RUTQVIST J., STEEFEL C., XU T., "Code inter-comparison builds confidence in numerical simulation models for geologic disposal of CO_2", *Energy*, vol. 29, pp. 1431–1444, 2004.

[RAJ 85] RAJAPAKSE R.K.N.D., SELVADURAI A.P.S., "On the performance of Mindlin plate elements in modelling plate-elastic medium interaction: a comparative study", *International Journal of Numerical Methods in Engineering,* vol. 23, pp. 1229–1244, 1985.

[REI 45] REISSNER E., "Effect of transverse shear deformation on bending of elastic plates", *Transactions of ASME, Journal of Applied Mechanics*, vol. 12, pp. A69–A77, 1945.

[RUT 02] RUTQVIST J., TSANG C.-F., "A study of caprock hydromechanical changes associated with CO_2 injection into a brine aquifer", *Environmental Geology*, vol. 42, pp. 296–305, 2002.

[RUT 07] RUTQVIST J., BIRKHOLZER J., CAPPA F., TSANG C.-F., "Estimating maximum sustainable injection pressure during geologic sequestration of CO_2 using coupled fluid flow and geomechanical fault-slip analysis", *Energy Conservation and Management*, vol. 48, pp. 1798–1807, 2007.

[RUT 08] RUTQVIST J., BIRKHOLZER J.T., TSANG C.-F., "Coupled reservoir-geomechanical analysis of the potential for tensile and shear failure associated with CO_2 injection in multilayered reservoir-cap rock systems", *International Journal of Rock Mechanics and Mining Sciences*, vol. 45, pp. 132–143, 2008.

[RUT 10] RUTQVIST J., VASCO D., MEYER L., "Coupled reservoir-geomechanical analysis of CO_2 injection and ground deformations at In Salah, Algeria", *International Journal of Greenhouse Gas Control*, vol. 4, pp. 225–230, 2010.

[SEL 79] SELVADURAI A.P.S., *Elastic Analysis of Soil-Foundation Interaction, Developments in Geotechnical Engineering*, vol. 17, Elsevier Scientific Publishing Co, The Netherlands, 1979.

[SEL 82] SELVADURAI A.P.S., "Axial displacement of a rigid elliptical disc inclusion embedded in a transversely isotropic elastic solid", *Mechanics Research Communications*, vol. 9, pp. 39–45, 1982.

[SEL 00a] SELVADURAI A.P.S., *Partial Differential Equations in Mechanics. II. The Bi-Harmonic Equation, Poisson's Equation*, Springer-Verlag, Berlin, 2000.

[SEL 00b] SELVADURAI A.P.S., *Partial Differential Equations in Mechanics. I. Fundamentals, Laplace's Equation, Diffusion Equation, the Wave Equation*, Springer-Verlag, Berlin, 2000.

[SEL 03a] SELVADURAI A.P.S., "Contaminant migration from an axisymmetric source in a porous medium", *Water Resources Research*, vol. 39, no. 8, pp. 1204, WRR 001742, 2003.

[SEL 03b] SELVADURAI A.P.S., "On intake shape factors for entry points in porous media with transversely isotropic hydraulic conductivity", *International Journal of Geomechanics*, vol. 3, pp. 152–159, 2003.

[SEL 04] SELVADURAI A.P.S., "Fluid intake cavities in stratified porous media", *Journal of Porous Media*, vol. 7, pp. 165–181, 2004.

[SEL 06] SELVADURAI A.P.S., "Gravity-driven advective transport during deep geological disposal of contaminants", *Geophysical Research Letters*, vol. 33, pp. L08408, 2006.

[SEL 07] SELVADURAI A.P.S., "The analytical method in geomechanics", *Applied Mechanics Reviews*, vol. 60, pp. 87–106, 2007.

[SEL 09] SELVADURAI A.P.S., "Heave of a surficial rock layer due to pressures generated by injected fluids", *Geophysical Research Letters*, vol. 36, pp. L14302, 2009.

[SEL 10] SELVADURAI A.P.S., SELVADURAI P.A., "Surface permeability tests: experiments and modelling for estimating effective permeability", *Proceedings of the Royal Society A: Mathematical, Physical and Engineering Sciences,* vol. 466, pp. 2819–2846, 2010.

[SEL 11] SELVADURAI A.P.S., "On the hydraulic intake shape factor for a circular opening located at an impervious boundary: influence of inclined stratification", *International Journal of Numerical and Analytical Methods in Geomechanics*, vol. 35, pp. 639–651, 2011.

[SEL 12a] SELVADURAI A.P.S., "Fluid leakage through fractures in an impervious caprock embedded between two geologic aquifers", *Advances in Water Resources*, vol. 41, pp. 76–83, 2012.

[SEL 12b] SELVADURAI A.P.S., "A geophysical application of an elastostatic contact problem", *Mathematics and Mechanics of Solids*, 2012.

[SNE 51] SNEDDON I.N., *Fourier Transforms*, McGraw-Hill, New York, 1951.

[TIM 59] TIMOSHENKO S.P., WOINOWSKY-KRIEGER S., *Theory of Plates and Shells*, McGraw-Hill, New York, 1959.

[VAS 08] VASCO D.W., FERRETTI A., NOVALI, F., "Reservoir monitoring and characterization using satellite geodetic data: interferometric synthetic aperture radar observations from the Krechba field, Algeria", *Geophysics*, vol. 73, pp. WA113–WA122, 2008.

[WIL 07] WILSON E., GERARD D. (eds), *Carbon Capture and Sequestration: Integrating Technology, Monitoring and Regulation*, Wiley-Blackwell, New York, 2007.

Chapter 6

Shear Behavior Evolution of a Fault due to Chemical Degradation of Roughness: Application to the Geological Storage of CO_2

CO_2 sequestration in deep geological formation is one way to reduce the greenhouse gas emissions causing global warming. Possible geological fault reactivation is an important aspect of site durability and must be considered. This phenomena stems from the modifications of fault balance, such as new effective stress or chemical deterioration of fault roughness. The chemical degradation of fault roughness defines the theme developed in this work: analysis of the fault roughness modifications induced by chemical degradation and their consequences on mechanical stability.

This chapter examines the evolution of the shear behavior of an existing crack in Campanian Flysch specimens with chemical degradation. For that purpose, crack surfaces were immersed in an acidic solution ([HCl] = 0.6 mol.L^{-1}) at a constant pH for 6 hours. A first analysis focused on the evolution of roughness parameters and material loss induced by chemical attack. Once degradation was characterized, direct shear tests on sound and chemically degraded samples were performed at the University of Sherbrooke (Québec, Canada). Classic curves of direct shear test (shear stress vs. tangential displacement and dilatancy) of damaged samples show a significant modification of the shear behavior. The degraded shear walls have no

---

Chapter written by Olivier NOUAILLETAS, Céline PERLOT, Christian LA BORDERIE, Baptiste ROUSSEAU and Gérard BALLIVY.

peak shear stress. Dilatancy curves show greater contractance for damaged joints. Results obtained from Barton's model are compared with experimental data.

6.1. Introduction

The greenhouse gas (GHG) emissions are a major cause of global warming [GEI 07]. CO_2 sequestration is a potential solution to reduce these emissions. Many projects around the world are performed to materialize theoretical research and to develop processes for its industrial development [ADE 05]. However, before the exploitation on a large scale, pilot projects have to guarantee the long-term safety of storage sites through the study of potential risks induced by CO_2 injection. One of these risks is seismic events caused by the reactivation of geological faults [BRG 08].

A geological fault can be considered simply as a discontinuity between two continuous media. There is slippage (or reactivation) when the shear stress acting on the fault exceeds the shear strength of the fault. In the context of CO_2 storage, this phenomenon stems from modifications of fault balance, due to a new state of effective stress [SOL 09, SOL 08, CAP 11] or the chemical deterioration of the fault roughness induced by the injection [EGE 06, BEM 09, OJA 11]. The last scenario can be the result of leakage through a pre-existing fracture in caprock. The gas dissolves in the underground water from the geological medium resulting in a low pH solution. This acidic flow could alter the properties of deep host formation. Most studies focused on the evolution of rock load bearing capacities (compressive strength, modulus of elasticity, etc.), to determine the modifications of the volumetric material characteristics. This chapter deals with the evolution of surface characteristics of the fault roughness. Indeed, the geochemical reactions, precipitation and dissolution, change the pre-existing fracture surface parameters (fault roughness, pore pressure, mineral composition of caprock, etc.) and could modify the equilibrium of the CO_2 storage site.

Typically, potential sites for CO_2 geological storage are located at great depth (>–5,000 m). The studied fault roughness is located at 2,500 m depth near mountains: the vertical stress is approximately 60 MPa and the horizontal stress is higher than 120 MPa. At those depths, stress and temperature conditions are enhanced and affect the rock's behavior. So, it is very important to take into account these parameters during experimental tests to simulate these particular *in situ* conditions. The literature presents a complete description of the effect of mechanical confinement on behavioral geomaterials. In these works, Mogi [MOG 07] shows that the rock, which is initially a brittle material, becomes ductile with confinement. In terms of mechanical properties, the modulus of elasticity increases with the confinement, especially as the rock is porous. The same observation can be expressed about the permanent strains and compressive strength. Consequently, confinement makes the rock more rigid in its elastic domain, more ductile in its plastic domain, and so more resistant. Unlike the effect of confinement, it is commonly accepted that the Young's modulus, the Poisson's ratio, the tensile strength, and the elasticity threshold decrease under the effect of temperature [HOM 86]. This decrease is mainly due to micro-cracking created by differential thermal expansion of minerals. These changes are predominant at 200 °C. In our case, as the temperature is lower than this value, we will neglect its effect.

Once the *in situ* conditions are represented directly or with effective stress, the experimental procedure comprises the chemical leaching of the crack's surface and the observation of the evolution of shear behavior. In contact with acid, the chemical species composing the rocks may dissolve at different degrees: a mineral composed of calcium, such as calcite, reacts more easily than quartz, which contains silica. The evolution of petrophysical properties, such as permeability and porosity, depends on the chemical reactions of minerals. The results supporting this point are numerous for carbonate rocks [EGE 06, BEM 09] or siliceous rocks [AND 08, LIU 03]. The increase in porosity for carbonate rocks is due to strong calcite dissolution which increases voids. This observation can also be extended to siliceous rocks, when the percentage of calcite is sufficient. However, a secondary material could precipitate and then fill the porous network and disconnect it. For geomechanical characteristics, experimental studies [EGE 06, BEM 09, OJA 11] have shown that the evolution of the isotropic elastic modulus, the shear modulus, the compressive strength, and the indirect tensile strength are related to the evolution of porosity, which is modified by chemical leaching. In general, the value of these parameters decreases with increasing porosity: this is explained by an increase of voids involving a decrease of the mechanical resistance of the surface. At the macroscopic scale and from a mechanical point of view, chemical degradation can be compared to damage. There are few references about the evolution of surface properties, especially roughness, after a chemical attack. Andreani [AND 08] focused on the crack opening due to an acidic drainage, but in that case, the sample was not submitted to *in situ* conditions.

The work presented subsequently aims to couple mechanical stress (shear) with a chemical attack, observing the surface evolution and the impact of the shear behavior. Then, experimental results are compared with those obtained from the Barton's model.

6.2. Experimental setup

Figure 6.1 shows the shear box used for our experimental test. Unlike a conventional setup, the normal force is provided by a horizontal hydraulic jack (6) (F_{max} = 50 kN) and the tangential force is measured by the load cell (3), which blocks the fixed half-block (1). The box is fixed on the jack of the MTS press (F_{max} = 3,500 kN), which is displacement controlled. This displacement causes the shearing of the mobile half-block (2) on the half-block which is fixed (1). Four

LVDT (5) are set around the diameter of the samples to measure the normal displacement (horizontal). One other LVDT (4) is placed above the shear box to observe the outgoing tangential displacement (vertical).

The characteristic curves of the shear test can be drawn from the measurements: tangential stress curve in terms of tangential displacement and the normal displacement curve in terms of tangential displacement (dilatancy).

A Campanian flysch from the caprock of the storage site of Lacq (France) was used as the testing material. This very impermeable sandstone is composed of more than 60% calcite. We have measured its mechanical and physical properties that are summarized in Table 6.1. The samples were taken from an outcrop, characteristic of Lacq storage material, and have been cored to a final diameter of 75 mm. A saw cut of 2 mm depth was made in the middle of the samples all around their diameter, and then they were broken in to two pieces by a three-point bending test. The saw cut enabled us to keep the crack remaining in the same plane.

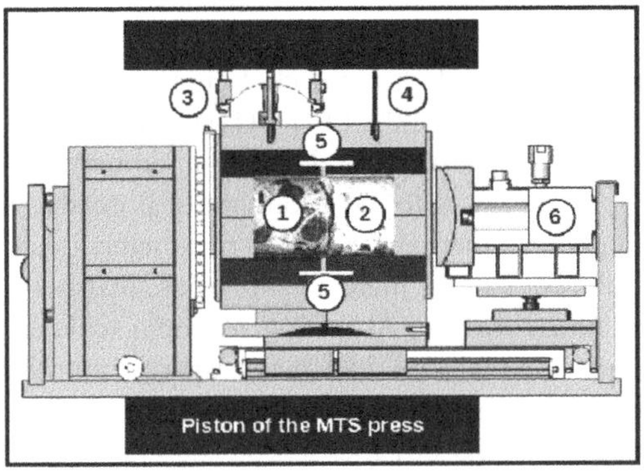

Figure 6.1. *Shear test box [ROU 10]*

Mechanical characteristics		Mean value	Standard deviation	Number of test
Compressive strength	Uniaxial test (MPa)	184	14	3
	Triaxial test (P_c = 0.15 MPa) (MPa)	197	7	3
Indirect tensile strength (MPa)		13	1	10
Bulk density (kg m^{-3})		2750	30	4
Water porosity (%)		0.28	0.02	4
Young's modulus (MPa)		56	3	8

Table 6.1. *Rock characterization*

6.3. Roughness and chemical attack

Many shear tests are well documented and are described in the literature. The originality of our test is the protocol for preparation of our samples. In order to observe the degradation of the roughness due to the leakage induced by the CO_2, three blocks were broken to give six shearing surfaces (the lower block is referred to as L and the upper block as U). They were immersed in an acidic solution ([HCl] = 0.6 mol L^{-1}, specimens referred as A) at a constant pH for 6 hours and compared to reference specimens immersed in water (referred as W). Since the natural leaching induced by the CO_2 dissolved in underground water is very slow, the degradation was accelerated using very low pH solutions. Scans of the roughness were performed before and after the chemical degradation with an accuracy of 0.1 mm to complete the visual examination. Figures 6.2(a) and (b) illustrate a wall scan before and after immersion in water. Figures 6.3(a) and (b) show the surface evolution of an attacked wall before and after the acid immersion. Visual examination and wall scan indicate a modification of joint geometry: the surface looks smoother. Smallest initial asperities disappear during chemical attack but new hollows are created due to the mineral heterogeneity.

a)

b)

Figure 6.2. *Wall scan of a reference half-block a) before water immersion and b) after water immersion*

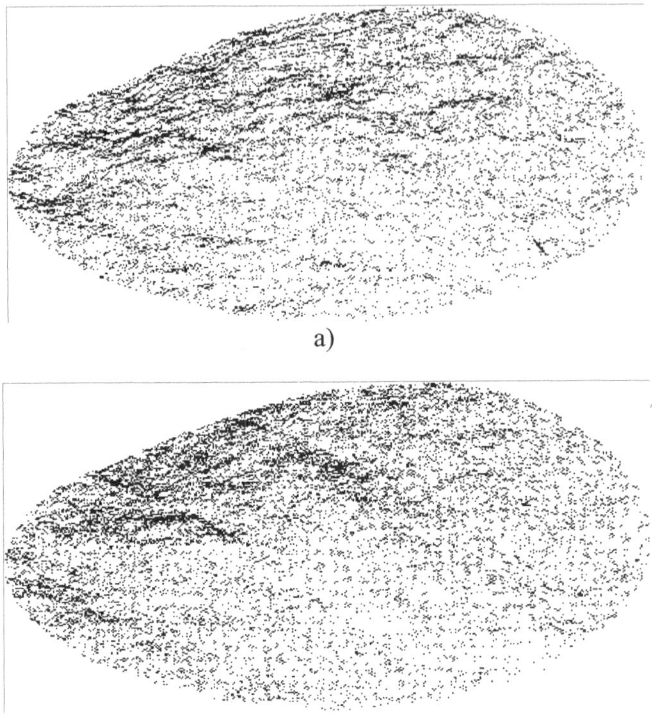

a)

b)

Figure 6.3. *Wall scan of a half-block a) before acid immersion and b) after acid immersion*

The numerical data allow the geometry of the surface crack to be reconstructed in future numerical analysis. To characterize the roughness of walls, the parameters *Z*2, *Z*3, *Z*4 and *RL* (defined in Table 6.2) were calculated (Δx = 0.5 mm) along the direction giving the lowest roughness that corresponds to the direction of the shear test. *Z*2 will be further used to calculate JRC in Barton's criterion.

Figure 6.4(a) shows two roughness parameters *Z*2 and *Z*3 obtained from the scan data of samples before immersion in water or in acidic solution. These values are low (*Z*2 < 0.5 and *Z*3 < 1) and have the same order of magnitude. So the roughness of samples is comparable and not pronounced.

Figure 6.4(b) presents the relative variation of *Z*2 induced by the immersion. This evolution for walls immersed in acid is not significant compared to those immersed in water.

The samples were weighed in air and in water to calculate their volume before and after the immersion. Figure 6.5 shows the material volume lost during the immersion. For the reference samples, volume change is negligible. The loss induced by immersion in acid is significant but it varies from one specimen to another. 6LA and 5LA are lower half-blocks that roughness was immersed in acid. As an example, the sample 6LA lost three times more material than the sample 5LA.

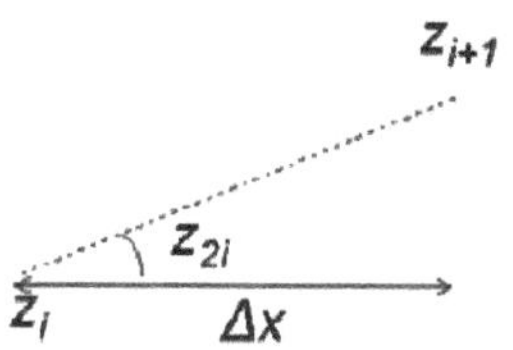	$Z2=\sqrt{\frac{1}{N}\sum_{i=1}^{n}\left(\frac{Z_{i+1}-Z_i}{\Delta x}\right)^2}$ [6.1]	– Similar to a topographic slope – The more the *Z2* is large, the more the roughness is pronounced
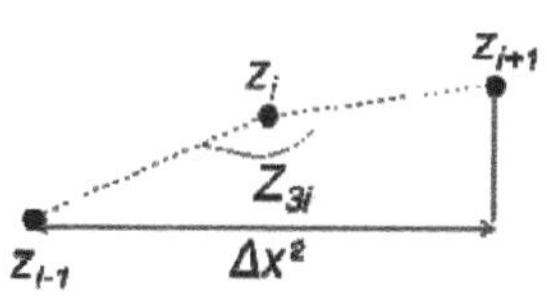	$Z3=\sqrt{\frac{1}{N}\sum_{i=1}^{n}\left(\frac{Z_{i-1}-2Z_i+Z_{i+1}}{\Delta x^2}\right)^2}$ [6.2]	– Comparable to a bending radius – The more the *Z3* is large, the more the roughness is pronounced
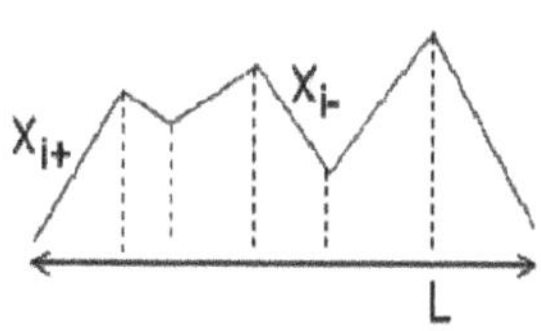	$Z4=\frac{\sum_{i=1}^{n}x_{i^+}-\sum_{i=1}^{n}x_{i^-}}{L}$ [6.3]	– Material mobilized during shearing – *Z4* is either positive or negative and depends on the direction of shear
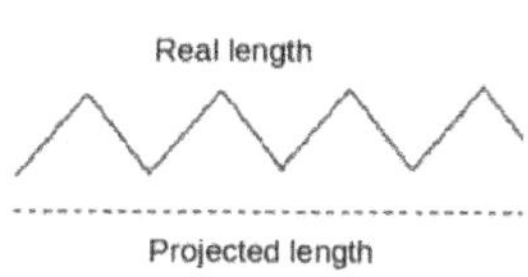	RL= Real length of profile / Length of the projection on the reference plane [6.4]	– The more the RL is, the more the roughness is pronounced

Table 6.2. *Roughness parameter definition [ROU 10]*

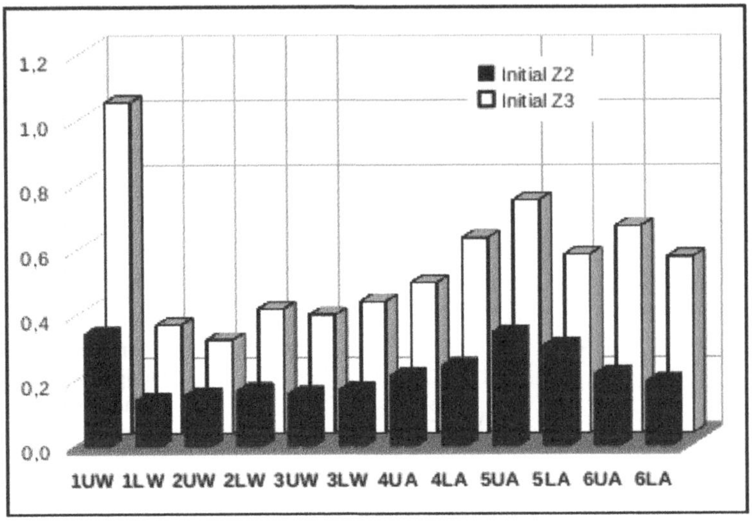

a)

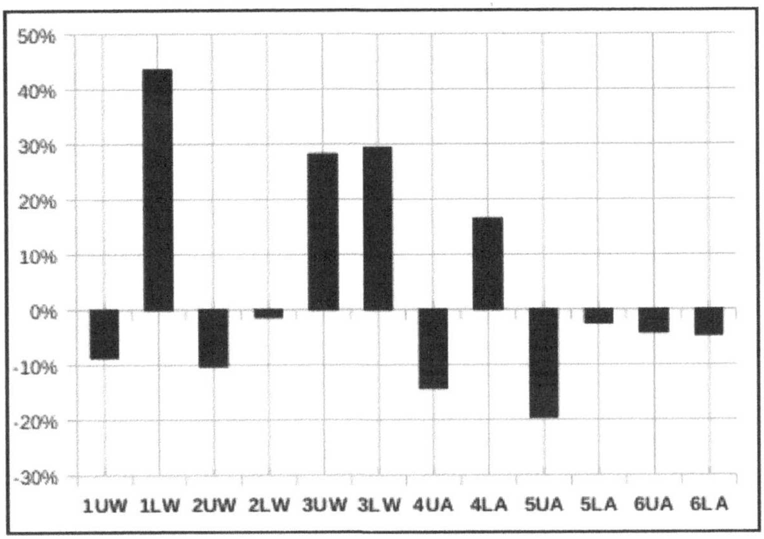

b)

Figure 6.4. *a) Z2 and Z3 roughness parameters before immersion; and b) Z2 roughness parameter evolution before and after immersion*

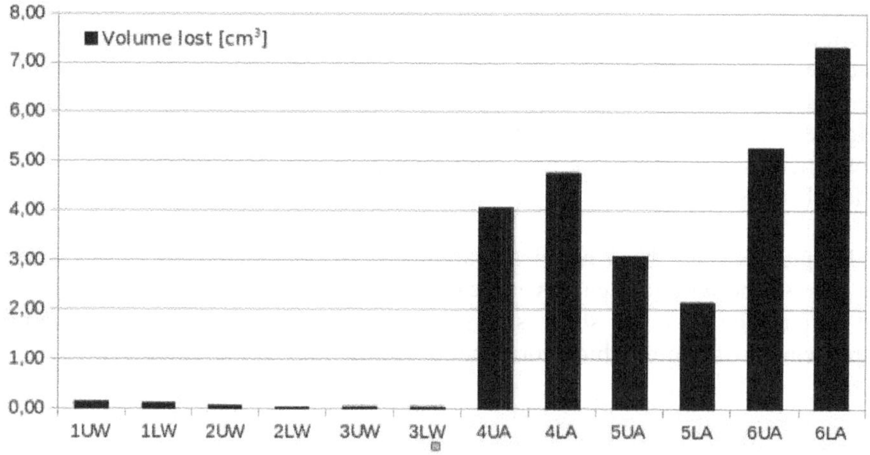

Figure 6.5. *Material volume lost during immersion*

This first part of the experiment does not permit any conclusion to be drawn on the modification of *Z2* parameter with chemical attack. Considering the surface heterogeneity, two antagonism phenomena could take place: on the one hand, the low initial roughness can be increased by the dissolution that creates hollows (Figure 6.3(b)). On the other hand, the larger asperities can be smoothed by the chemical attack. However, these two phenomena impact profiles of both walls of the discontinuity, which eventually no longer match. Despite these observations on the roughness parameters, the hardness of the material is modified by the chemical attack, and a modification on the shear behavior of the specimens is expected.

6.4. Shear tests

Direct shear tests are performed at constant force stress of 4.42 kN which corresponds to an initial normal stress of 1 MPa, a tangential displacement rate of 0.15 mm/min is imposed until the maximal value of 6 mm. A low normal stress is chosen to highlight the effect of roughness.

Figure 6.6 illustrates typical curves of shear stress versus tangential displacement for high and low roughness, smooth surface [LOP 00], and Figure 6.7 illustrates our experimental results. The shear and normal stresses are the ratio between the corresponding loads and the mean contact surface area which is corrected by the measured shear displacement.

For specimens immersed in water and specimens immersed in acidic solution, the average behavior is calculated from the results of three tests. These specimens are surrounded by the extrema values. In comparison to Figure 6.6, samples immersed in water present a behavior that corresponds to the low roughness case, as suggested by the calculated roughness parameters: shear stress increases quickly to a peak and then decreases to a plateau equal to the residual stress.

Despite the fact that roughness parameters of degraded samples are similar, the corresponding shear curves indicate that they behave like smooth samples. We can notice that the shear stress continues to increase after a tangential displacement of 6 mm, which corresponds to the limit of our equipment.

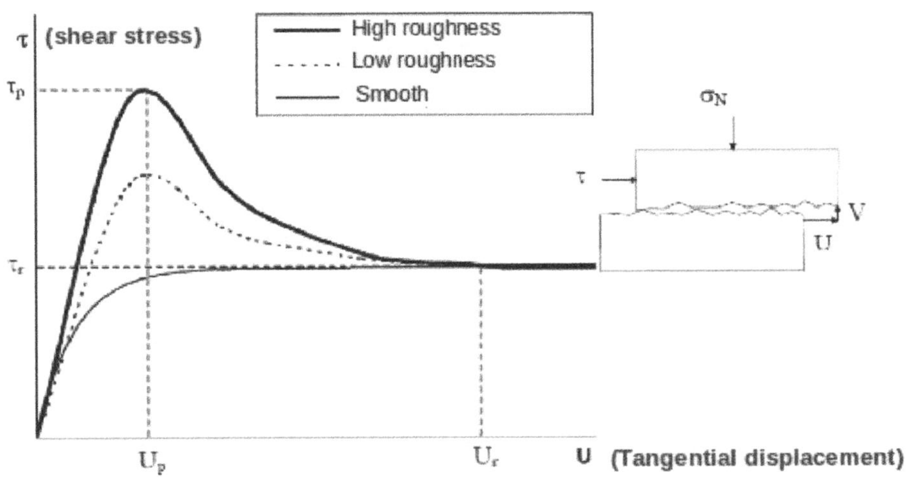

Figure 6.6. *Classical curves of shear stress versus tangential displacement [LOP 00]*

Figure 6.8 represents the dilatancy curves and their extrema variations in terms of the normal stress. Typically, the dilatancy illustrates crack aperture variation (*V*) with the shear displacement (*U*) (Figure 6.8). At the beginning of the test, the crack opening decreases – this is the contractancy that corresponds for the reference specimens to the crack's return to its original position – then the crack opening increases first linearly and smoothly and then tends to plateau value which depends on the normal stress. Experimental results show (Figure 6.9) that the contractancy of degraded samples is more significant than in reference samples.

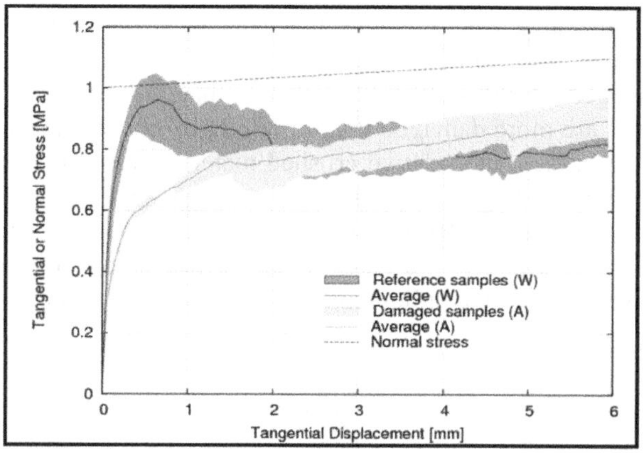

Figure 6.7. *Experimental curves of tangential stress versus tangential displacement*

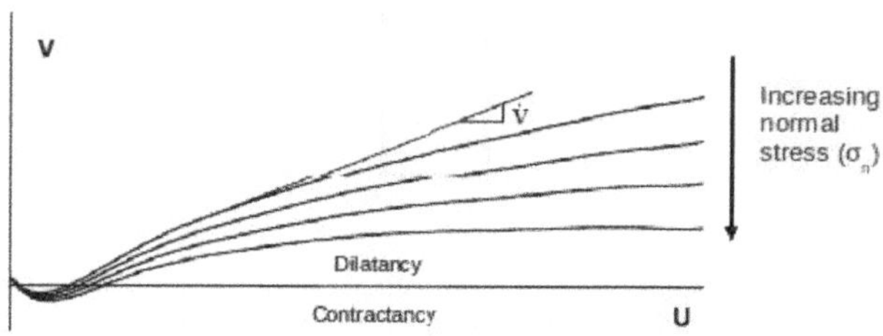

Figure 6.8. *Classical curves of dilatancy [LOP 00]*

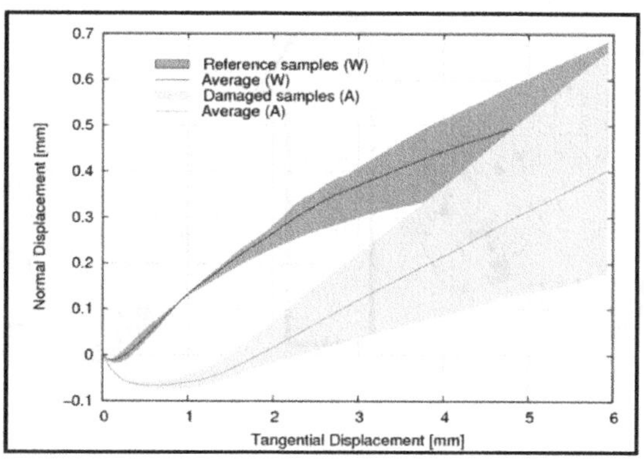

Figure 6.9. *Experimental curves of dilatancy*

Figures 6.10(a) and (b) illustrate the contact areas after the shear test. Mechanical damage is located at the more clear areas for both reference and chemically damaged samples. We can observe that the asperities of chemically degraded samples are more damaged than in reference samples for the same normal load. As a first approach, the relative crushed material area can be estimated from image analysis (thresholding and binarization) of Figures 6.11(a) and (b); the values evaluated are 9% for the reference sample and 20% for the attacked sample. Obviously, this result can only be taken as qualitative as the original color of the samples and the corresponding enlightenments were not controlled.

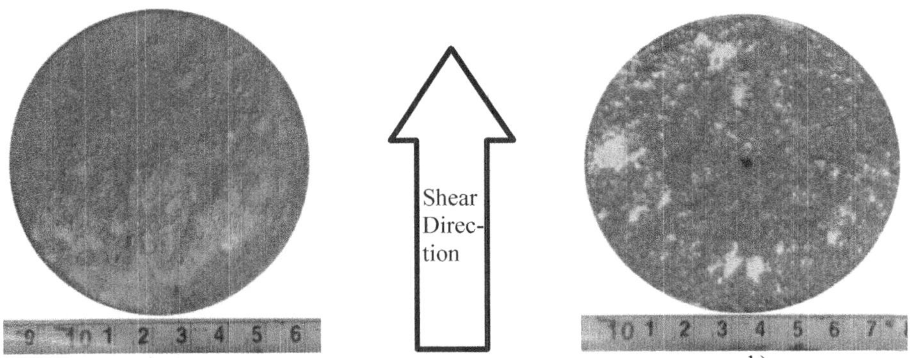

a) b)

Figure 6.10. *a) Roughness of a reference surface after shear test; and b) roughness of a damaged surface after shear test*

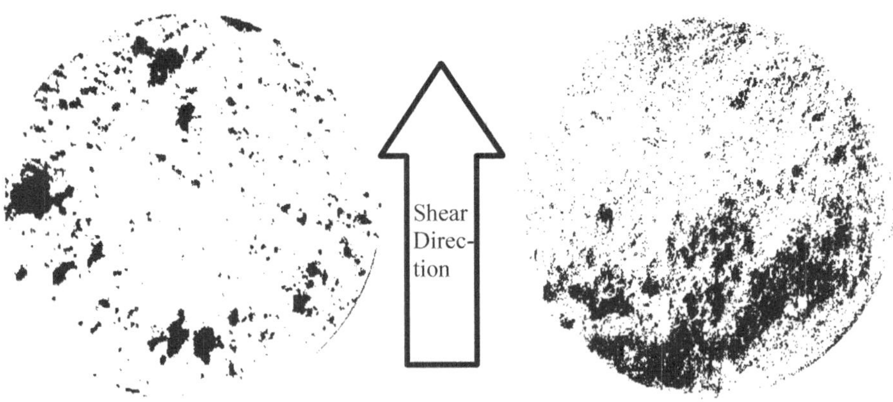

a) b)

Figure 6.11. *a) Crushed material area of a reference surface; and b) crushed material area of a damaged surface*

These results reveal that the chemical degradation of the crack surface causes modification in shear behavior with the disappearance of the peak shear stress, the increasing of contractancy, and an enhancement of the damage area after shearing. These evolutions are certainly due to the wall geometry mismatch and the reduction hardness of degraded surfaces, even if there is no significant modification of the roughness parameter.

The area between the two curves in Figure 6.7 could be interpreted as a loss of fracture energy that could potentially have been mobilized during the test.

The increase of contractancy (Figure 6.9) is the result of the fact that the wall geometry of degraded samples no longer corresponds to the modification of the crack surface resulting from the chemical attack. The dilatancy curve of chemically damaged samples reaches its minimum when the shear stress versus tangential displacement curve grows more gradually. The initial sliding is accompanied by a destruction of the weakened layer of chemically damaged material, and then the slip becomes predominant compared to mechanical damage. The dilatancy curve becomes linear as for a standard shear test, but with a greater average slope. The hollow on the surface will therefore fill more quickly, resulting in an acceleration of the opening.

On the other hand, the residual shear strength of degraded samples (Figure 6.7) is greater than the shear strength of reference samples. This could be explained by the increase of the volume of sheared material.

This experimental campaign reveals the influence of chemical surface degradation on the shear behavior of rock joints. From this statement, we consider Barton's model to observe its capacity to take into account wall degradation.

6.5. Peak shear strength and peak shear displacement: Barton's model

Predicting shear behavior of a rock joint has been a recurrent problem for the last 40 years. Development of underground sequestration of CO_2 and radioactive waste disposal increases the importance of knowing the shear strength to ensure the sustainability of sites. Several empirical and theoretical constitutive models were proposed to determine these criteria separating pre-sliding and post-sliding of a joint [GRA 03] and to simulate the shear behavior of a fracture. Among them, Barton's model is the most commonly used [GRA 03, IRS 78] due to its simplicity and because it takes into account roughness parameters of the crack. Its accuracy to determine the shear strength is acceptable for sound materials but must be evaluated for chemically degraded material.

Barton and Choubey [BAR 77] proposed the following expression for the shear strength of a rock joint:

$$\tau = \sigma_n \tan\left(JRC * \log_{10}\left(\frac{JCS}{\sigma_n}\right) + \varphi_r \right) \qquad [6.5]$$

where:

- τ is the peak shear stress;
- σ_n is the effective normal stress;
- JRC is the joint roughness coefficient;
- JCS is the joint wall compressive strength;
- φ_r is the residual fraction angle.

The value of the effective normal stress (σ_n) is the ratio between the normal force applied (F_n = 4.42 kN) and the shearing surface. It should be constant. In our tests, the normal force is constant, but the shearing surface evolves. Knowing the value of the tangential displacement, we can calculate the surface contact at any time and σ_n is taken from experiments at the peak shear stress (Table 6.3).

Barton and Choubey estimate the joint roughness coefficient (JRC) from visual comparison with the 10 standard profiles [BAR 77]. This method is prone to subjectivity. To determine JRC, Tse and Cruden [TSE 79] developed an empirical statistical relationship between the JRC and the roughness parameter Z_2 (Table 6.2):

$$JRC = 32.2 + 32.47\log_{10}(Z_2) \qquad [6.6]$$

JRC was calculated for both walls of the joint according to equation [6.6] and the mean value was introduced in Barton's equation (Table 6.3).

Sample	1W	2W	3W	4A	5A	6A
τ_{peak} (MPa)	1.05	0.87	1.03	0.98	0.85	0.88
δ_{peak} (mm)	0.63	0.29	0.75	5.86	6.00	6.00
σ_n (MPa)	1.08	1.03	1.02	1.11	1.12	1.09
JRC	13.64	6.44	11.12	12.26	15.09	9.97

Table 6.3. *Experimental values*

Values of joint wall compressive strength (JCS) are taken as equal to the compressive strength (σ_c) in unweathered rock. When the walls are damaged, JCS can be determined with Schmidt hammer tests, but a conservative lower limit is allowed in the absence of this test [IRS 78]: $JRC = \frac{1}{4}\sigma_c$.

In our case, values of JCS vary between each extremum:

$$\frac{1}{4}\sigma_c = 46 \text{ MPa} < JCS < \sigma_c = 184 \text{ MPa}$$

The residual friction angle of the rock was determined with a series of triaxial tests at different confining pressures: 0.3, 20, 40 and 60 MPa. From these results, the slope of the tangential line at Mohr Coulomb's circles was determined (Figure 6.12).

$$\varphi_r = 37°$$

Figure 6.13 presents results of Barton's criterion in terms of experimental results. The values of the model overestimate the experimental values by a factor of 1.77 for the reference samples and 2.31 for the damaged samples.

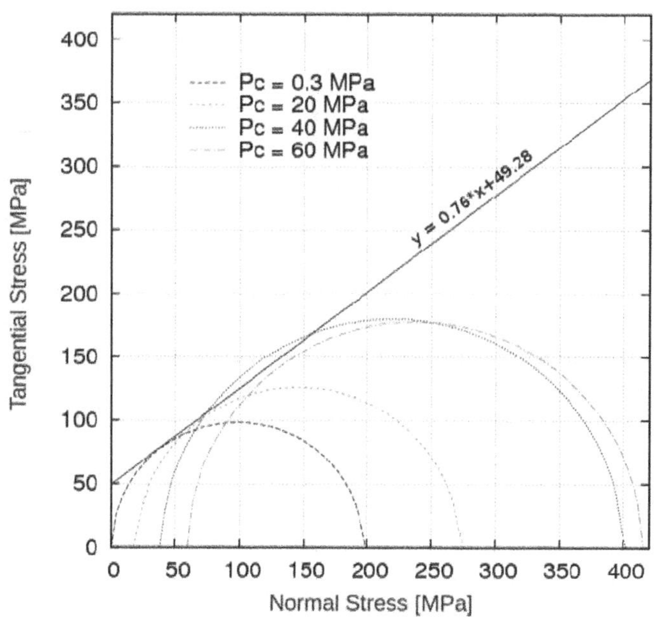

Figure 6.12. *Mohr Coulomb's circles resulting from the triaxial tests*

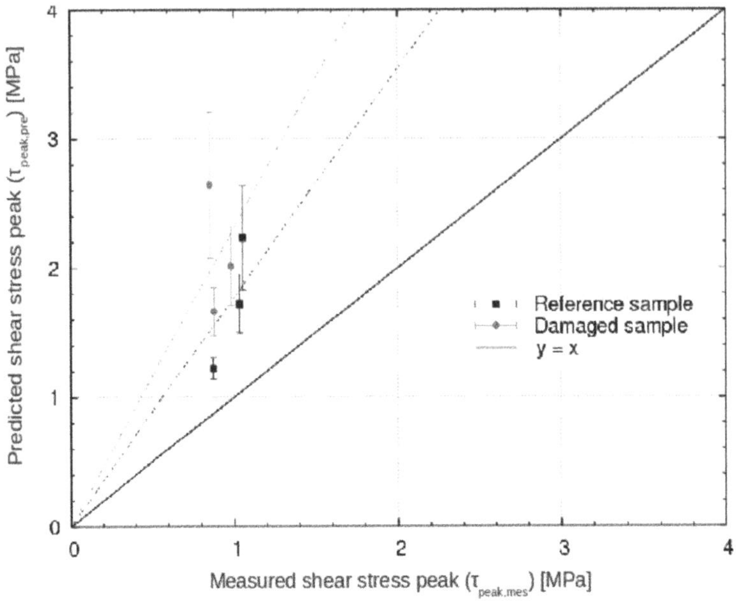

Figure 6.13. *Peak shear stress: Barton's criterion versus experimental result*

To simulate shear behavior of a rock joint, another datum is extremely important: the shear stiffness [BAR 77, TSE 79]. Secant peak shear stiffness (K_s) is the ratio between the value of peak shear stress (τ_{peak}, MPa) and the value of the peak tangential displacement (δ_{peak}, mm).

$$K_s = \frac{\tau_{peak}}{\delta_{peak}} \qquad [6.7]$$

To determine the peak shear displacement of rock joints, Barton and Choubey [BAR 77] suggested the following empirical equation:

$$\frac{\delta_{peak}}{L} = \frac{1}{500} * \left(\frac{JRC}{L}\right)^{0.33} \qquad [6.8]$$

where L is the length of jpoint sample (m).

This equation is a limit for smooth roughness, when the JRC is equal to 0. Indeed, in this case, peak shear stress is reached for zero displacement. To overcome these limitations, Asadollahi *et al.* [ASA 10] introduced the following empirical equation:

$$\delta_{peak} = 0.0044 * L^{0.34} * \left(\frac{\sigma_n}{JCS}\right) * \cos\left(JRC \log_{10}\left(\frac{JCS}{\sigma_n}\right)\right) \qquad [6.9]$$

An experimental validation [ASA 10] gives this new equation but only for a rough fracture of sandstone, whereas the original Barton's equation is validated in granite and limestone joints.

We calculate the peak shear displacement with both equations [6.8] and [6.9]. Figure 6.14 presents the results of equations in terms of the experimental peak shear displacement.

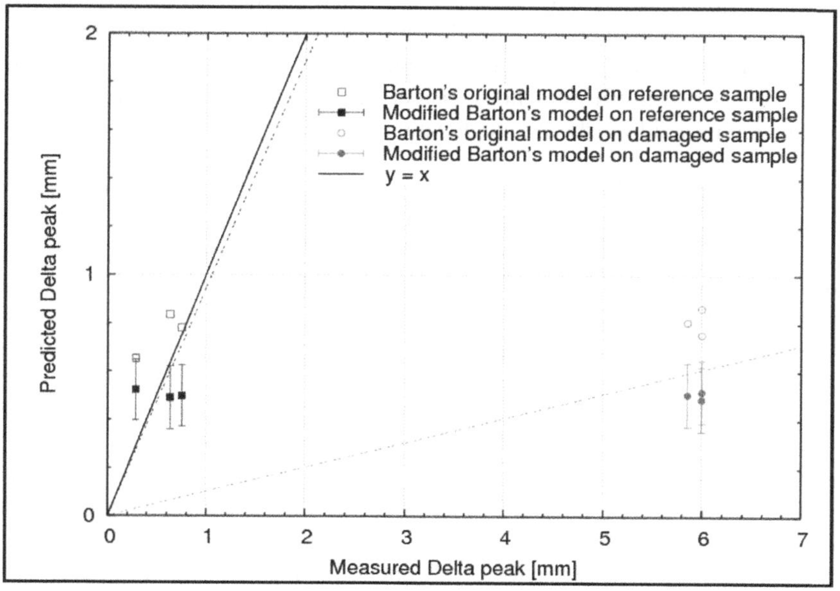

Figure 6.14. *Peak shear displacement: Barton's criterion versus experimental results*

The straight line of equation $y = x$ from Figure 6.14 indicates that the models have been validated for reference samples. For damaged samples, the experimental results diverge from the models' validation line. The original Barton's model and the modified model do not consider the aperture of the joint. This parameter is important especially for damaged samples that have the initial fitting of walls modified by the chemical attack. It will be interesting to compare this result with that of Ladany–Archambault model that considers the joint's aperture.

6.6. Conclusion and perspectives

While a first analysis revealed that the acid attack physically modifies crack surfaces, the evolution of roughness parameter Z2 is difficult to analyze due to the heterogeneity of rock walls. However, calculation of the dissolved material mass confirms the visual examination: there is an alteration of wall geometry. Consequently, the initial fitting of half-blocks is modified after the chemical degradation. These changes are clearly seen with mechanical testing: Shear behavior of degraded joints is different. Despite an apparent roughness of walls, the shear stress versus tangential displacement curves of degraded samples illustrate the behavior of a smooth joint. On the dilatancy curve, the contractancy is more significant for the degraded joints than the reference joints. This increase can be explained by the wall geometry mismatch and the reduction in hardness of degraded surfaces. These results mean that Barton's model could not be used to calculate displacement at the shear stress peak of the damaged joints because it does not consider the crack aperture. However, Barton's criterion remains valid for shear strength values.

The limitation of this test is the setup scale. Tests in the laboratory cannot reproduce stress *in situ*, we have to work with effective stress. In addition, the leaching does not occur at the same time as the shear, which does not represent the flow of acid fluid into the crack. This experimental approach also simplifies the real problem. Indeed, we represent a fault with a crack; however, the fault-zone structure is more complicated. In general, three areas were distinguished: the core zone (thickness between 1 and 10 cm), the damaged zone (between 1 and 100 m) and the intact zone. The first zone determines the sliding area, containing gouge or not. The damaged zone is composed of fractured rock, and the last zone represents the sound material. We cannot quantify damage caused by a three-point bending test. The direct shear test is 2D sliding: it does not take into account the fact that intermediate stress plays an important role in the behavior of the rock. Finally, these tests have shown a modification in the shear behavior of a joint following chemical degradation of its walls, but they do not simulate the case of a CO2 leakage along a fault.

In conclusion, this first experimental approach to the problem is initially satisfactory because it simplifies the phenomena. Separating the chemical attack and mechanical stress will be very useful for modeling through the scans performed throughout the protocol. However, if we wish to reproduce the phenomenon *in situ*, the test quickly shows its limitations. In the first place, to resolve this problem, a triaxial test with acid flow in a crack is being developed. Then, experimental campaigns will allow a numerical model to be validated via simulations of these laboratory tests. Through an upscaling law, the modelization of the phenomenon can be represented at the *in situ* scale.

6.7. Acknowledgment

The authors would like to thank the LMRGA laboratory of the University of Sherbrooke and especially Mr. Lalonde, the senior technician, who ensured the experiments were carried out successfully.

6.8. Bibliography

[ADE 05] ADEME, IFP, BRGM, La capture et le stockage géologique du CO_2, 2005.

[AND 08] ANDREANI M., GOUZE P., LUQUOT L., JOUANNA P., "Changes in seal capacity of fractured claystone caprocks induced by dissolved and gaseous CO_2 seepage", *Geophysical Research Letters*, vol. 35, L14404, 6 pp, 2008.

[ASA 10] ASADOLLAHI P., INVERNIZZI M.C.A., ADDOTTO S., TONON F., "Experimental validation of modified Barton's Model for rock fractures", *Rock Mechanics*, vol. 43, pp. 597–613, 2010.

[BAR 77] BARTON N., CHOUBEY V., "The shear strength of rock joints in theory and practice", *Rock Mechanics*, vol. 10, pp. 1–54, 1977.

[BEM 09] BEMER E., LOMBARD J.M., "From injectivity to integrity studies of CO_2, geological storage", *Oil & Gas Science and Technology – Revue de l'Institut Français du Pétrole*, vol. 65, no. 3, pp. 445–459, July, 2009.

[BRG 08] BRGM, Rapport final du dossier 'Code Minier' pour l'autorisation du pilote de stockage géologique de CO_2 à Rousse (64), 2008.

[CAP 11] CAPPA F., RUTQVIST J., "Modeling of coupled deformation and permeability evolution during fault reactivation induced by deep underground injection of CO_2", *International Journal of Greenhouse Gas Control*, vol. 5, no. 2, pp. 336–346, March, 2011.

[EGE 06] EGERMANN P., BEMER E., ZINSZMER B., "An experimental investigation of the rock properties evolution associated to different levels of CO_2 injection like alteration process", *International Symposium of the Society of Core Analysts*, Trondheim, Norway, 2006.

[GEI 07] GROUPE D'EXPERT INTERGOUVERNEMENTAL SUR L'ÉVOLUTION DU CLIMAT, Changements climatiques, 2007.

[GRA 03] GRASSELLI G., EGGER P., "Constitutive law for the shear strength of rock joints based on three-dimensional parameters", *International Journal of Rock Mechanics and Mining Sciences*, vol. 40, pp. 25–40, 2003.

[HOM 86] HOMAND F., *Comportement mécanique des roches en fonction de la température*, Éd. de la Fondation scientifique de la géologie et de ses applications, Vandoeuvre-les-Nancy, 1986.

[IRS 78] IRSM, "Suggested methods for the quantitative description of discontinuities in rock masses", *International Journal of Rock Mechanics and Mining Sciences & Geomechanics Abstracts*, vol. 15, pp. 319–368, 1978.

[LIU 03] LIU L., SUTO Y., BIGNALL G., YAMASAKI N., HASHIDA T., "CO_2 injection to granite and sandstone in experimental rock/hot water systems", *Fuel and Energy Abstracts*, vol. 44, no. 5, p. 310, 2003.

[LOP 00] LOPEZ P., Comportement mécanique d'une fracture en cisaillement: Analyse par plan d'expériences des données mécaniques et morphologiques connues sur une fracture, University of Bordeaux I and University of Québec at Chicoutimi, 2000.

[MOG 07] MOGI K., *Experimental Rock Mechanics*, Taylor & Francis, New York, London, 2007.

[OJA 11] OJALA I.O., "The effect of CO_2 on the mechanical properties of reservoir and cap rock", *Energy Procedia*, vol. 4, pp. 5392–5397, Jan. 2011.

[ROU 10] ROUSSEAU B., "Interfaces fragiles des ouvrages hydrauliques: Morphologie et comportement mécanique", University of Bordeaux 1 and University of Sherbrooke, 2010.

[SOL 08] SOLTANZADEH H., HAWKES C.D., "Semi-analytical models for stress change and fault reactivation induced by reservoir production and injection", *Journal of Petroleum Science and Engineering*, vol. 60, no. 2, pp. 71–85, Feb. 2008.

[SOL 09] SOLTANZADEH H., HAWKES C.D., "Assessing fault reactivation tendency within and surrounding porous reservoirs during fluid production or injection", *International Journal of Rock Mechanics and Mining Sciences*, vol. 46, no. 1, pp. 1–7, Jan. 2009.

[TSE 79] TSE R., CRUDDEN D.M., "Estimating joint roughness coefficients", *International Journal of Rock Mechanics and Mining Sciences*, Abstract vol. 16, Issue 5, pp. 303–307, 1979.

Chapter 7

CO_2 Storage in Coal Seams: Coupling Surface Adsorption and Strain

7.1. Introduction

About 10% of the natural gas produced in the United States is coal bed methane (CBM), i.e. methane extracted from unmineable coal deposits. The amount of this gas recovered from a reservoir can be increased by injecting carbon dioxide into the reservoir, a process known as enhanced coal bed methane recovery (ECBM) [WHI 05]. Indeed, carbon dioxide is preferentially adsorbed on coal with respect to methane, and an injection of the former will therefore ease the recovery of the latter from the coal matrix. Moreover, the injected carbon dioxide can potentially remain stored in the reservoir over geological periods of time, thus lowering the carbon footprint associated with the consumption of the produced methane. However, large-scale pilot projects of ECBM have demonstrated that the injection of carbon dioxide leads to a decrease of the injectivity of the reservoir (see, e.g. the Allison pilot project in the United States [REE 03]), which impedes the economic viability of such projects.

The depth of coal bed reservoirs typically ranges from 300 m to about 1.5 km. These reservoirs are naturally fractured: it is this set of natural fractures (called cleats), spaced a few centimeters from each other, which govern the transport properties of the reservoir. In between those cleats, we find the coal material itself. Coal is a porous medium, with pores ranging from the sub-nanometric size up to the

---

Chapter written by Saeid NIKOOSOKHAN, Laurent BROCHARD, Matthieu VANDAMME, Patrick DANGLA, Roland J.-M. PELLENQ, Brice LECAMPION and Teddy FEN-CHONG.

micrometric size. Coal is known to swell in the presence of carbon dioxide or methane. At a given pressure of the fluid, this swelling is more pronounced with carbon dioxide than with methane [LEV 96]. Therefore, an injection of carbon dioxide into a reservoir initially full of methane results in a differential swelling of the coal matrix. In the confined conditions which prevail underground, this differential swelling leads to a closure of the cleat system, which translates into a decrease in the injectivity of the reservoir. The evolution of the injectivity of the reservoir during the process of injection is therefore the result of a coupling between adsorption, swelling of the coal matrix, and transport properties of the reservoir. Recently, several models of reservoir have been proposed to capture this coupling [ZHA 08, CHE 10, WU 10, WU 11].

In this chapter, we propose a multi-scale approach in order to estimate the complex evolutions of injectivity over the process of injection. This approach combines molecular simulations and reservoir simulations. At the scale of the reservoir, we derive a model for the coal-bed reservoir based on the Biot–Coussy poromechanical framework [COU 04]. Two porous networks are explicitly taken into account: the cleats of the reservoir and the pores of the coal material. The model uses poromechanical equations that have been extended recently to the effect of surface adsorption [COU 10, VAN 10]. The resulting model requires directly as an input the adsorption isotherm of the fluid on coal. We show how this adsorption isotherm can be obtained by molecular simulations.

In the first section, the poromechanical model for the coal bed reservoir is derived. In the second section, molecular simulations are performed in order to estimate the adsorption isotherm of carbon dioxide on coal. The simulated adsorption isotherm is then used as an input for reservoir simulations of injection of carbon dioxide in a coal bed reservoir.

7.2. Poromechanical model for coal bed reservoir

This section is devoted to the derivation of the constitutive equations which govern the behavior of a coal bed reservoir during an injection of fluid. In order to give the model a physical basis, we start by explaining why coal swells in the presence of carbon dioxide.

7.2.1. *Physics of adsorption-induced swelling of coal*

Let us consider a soap bubble. As was shown early on by Laplace [LAP 06] and Young [YOU 05], creating surface requires providing energy to the system, from what follows the pressure inside the bubble differs from the pressure outside the bubble:

$$p_{in} - p_{out} = 2\tilde{\sigma}^s / R \qquad [7.1]$$

where R is the radius of the bubble, $\tilde{\sigma}^s$ is the surface stress, and p_{in} and p_{out} are the pressures inside and outside of the bubble, respectively. By mechanical analogy, a soap bubble behaves like a stretched membrane. Then, Gibbs showed that adsorption – an accumulation of molecules of fluid on an interface – can modify the surface stress. More precisely, the well-known Gibbs law [GIB 28] for fluid–fluid interfaces shows that adsorption leads to a decrease of surface stress:

$$d\tilde{\sigma}^s = -\Gamma d\mu \qquad [7.2]$$

where μ is the molar chemical potential of the adsorbed molecules of fluid and Γ is the molar amount of adsorbed molecules of fluid in excess of the bulk density per unit area of interface. Combining equations [7.1] and [7.2], we find that adsorption leads to a decrease of the pressure inside the soap bubble, which can only be achieved by a swelling of the bubble.

Likewise, in coal, the surface of the pores is an interface between a solid (the solid skeleton) and a fluid (e.g. carbon dioxide). Therefore, the surface of the pores also behaves like a stretched membrane. Adsorption of fluid at the surface of the pore will modify how this membrane is stretched, which will lead to a deformation of the coal sample.

Energy can be provided to the surface of the pores either by straining this surface (and working against the surface stress) or by adding fluid molecules on the surface. The energy balance for the surface of the pores is therefore:

$$df_{surf} = \tilde{\sigma}^s ds_p + \mu_p dn_p^a \qquad [7.3]$$

where f_{surf} is the Helmholtz free energy of the pore surface per unit volume of undeformed porous medium, s_p is the surface of the pores per unit volume of undeformed porous medium, μ_p is the molar chemical potential of the fluid in the coal pores, and n_p^a is the molar density of fluid molecules adsorbed at the surface of the pores per unit volume of undeformed porous medium.

The pore surface is an interface between a solid and a liquid. For such an interface, if the adsorption does not depend on the strain of the surface, Gibbs' law [7.2], which was initially derived for an interface between two fluids, remains valid [SHU 50]. In an integrated form, Gibbs' law can be written as follows:

$$\Delta\tilde{\sigma}^s = -\int_{p=0}^{p} \Gamma(\mu) d\mu = -\int_{p=0}^{p} \Gamma(p)\bar{V}_b(p) dp \qquad [7.4]$$

where p and $\overline{V}_b(p)$ are the bulk pressure and the bulk molar volume of the fluid, respectively.

7.2.2. *Assumptions of model for coal bed reservoir*

The coal bed reservoir is modeled as a porous medium made of two types of porosity: cleats and coal porosity (see Figure 7.1).

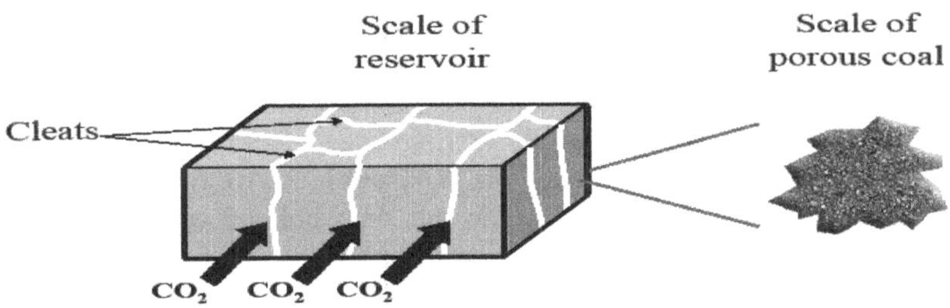

Figure 7.1. *Model for coal bed reservoir*

The transport properties are assumed to be governed by the cleat system only. In contrast, adsorption is assumed to occur at the surface of the coal pores only: swelling is due to the coal material. Therefore, fluid molecules can be found in a bulk state in the cleats or in a bulk state in the coal pores or in an adsorbed state at the surface of the coal pores.

We assume that the characteristic time of transfer of fluid from the cleat system to the coal porosity and from the coal porosity to the cleat system is much smaller than any other characteristic time of transport of fluid in the reservoir: at each location in the reservoir, we assume that the chemical potential of the fluid in the cleats is equal to the chemical potential of the fluid in the coal pores.

We consider nonlinear (i.e. stress-dependent) poroelastic properties of the reservoir. For the sake of simplicity, only the case of a pure component (i.e. carbon dioxide only) is considered.

7.2.3. *Case of coal bed reservoir with no adsorption*

For now, the fluid pressure p_c in the cleats is considered to differ from the fluid pressure p_p in the coal pores. We consider a representative elementary volume

(REV) of the reservoir and note f_{skel} the Helmholtz free energy of its solid skeleton per unit volume of undeformed medium. If no adsorption occurs in the reservoir, the reservoir is a classical dual-porosity medium. For such a system, the energy balance for the solid skeleton is [COU 04]:

$$df_{skel} = \sigma d\varepsilon + s_{ij} de_{ij} + p_c d\phi_c + p_p d\phi_p \qquad [7.5]$$

where σ is the volumetric confining stress, ε is the volumetric strain, s_{ij} are the deviatoric confining stresses, e_{ij} are the deviatoric strains, ϕ_c is the Lagrangian porosity associated to the cleats (i.e. the volume fraction of the cleats with respect to the REV), and ϕ_p is the Lagrangian coal porosity.

For such a nonlinear dual-porosity medium, the constitutive equations in an incremental form are [COU 04]:

$$\begin{cases} d\sigma = Kd\varepsilon - b_c dp_c - b_p dp_p \\ ds_{ij} = 2Gde_{ij} \\ d\phi_c = b_c d\varepsilon + dp_c / N_c + (dp_c - dp_p) / N_{cp} \\ d\phi_p = b_p d\varepsilon + dp_p / N_p + (dp_p - dp_c) / N_{cp} \end{cases} \qquad [7.6]$$

where the bulk modulus K, the shear modulus G, and the other poroelastic parameters b_c, b_p, N_c, N_c, and N_{cp} of the reservoir all depend on the confining stress σ and on the fluid pressures p_c and p_p.

If we apply $p_c = p_p$, we must retrieve the classical constitutive equations of a porous medium with a single porosity. For such a medium, the solid phase is the solid skeleton, whose bulk modulus is noted to be K_s. Some classical relations exist, which link the different poroelastic parameters [COU 04]:

$$\begin{cases} b_c + b_p = 1 - K / K_s \\ 1 / N_c = (b_c - \phi_c) / K_s \\ 1 / N_p = (b_p - \phi_p) / K_s \end{cases} \qquad [7.7]$$

In contrast, if we apply a pressure p_c in the cleats while keeping a zero pressure $p_p = 0$ in the coal pores, we must again retrieve the classical constitutive equations of a porous medium with a single porosity. But, for such a medium, the solid phase is now the coal material itself, whose bulk modulus is noted to be K_m. The classical relations used to derive equation [7.7] now yield:

$$\begin{cases} b_c = 1 - K / K_m \\ 1/N_c + 1/N_{cp} = (b_c - \phi_c)/K_m \end{cases} \quad [7.8]$$

7.2.4. *Derivation of constitutive equations for coal bed reservoir with adsorption*

We consider a representative elementary volume (REV) of the reservoir and note f its Helmholtz free energy per unit volume of undeformed material. The fluid pressure p_c in the cleats is still considered to differ from the fluid pressure p_p in the coal pores: those two pressures will only be equated at the end of the derivation. Energy can be provided to the REV of the reservoir by straining it, by adding molecules of fluid in the cleats, or by adding molecules of fluid in the coal pores:

$$df = \sigma d\varepsilon + s_{ij} de_{ij} + \mu_c dn_c + \mu_p dn_p \quad [7.9]$$

where μ_c is the molar chemical potential of the molecules of fluid in the cleats, n_c is the molar density of fluid molecules in the cleats per unit volume of undeformed porous medium, μ_p is the molar chemical potential of the molecules of fluid in the coal pores, and n_p is the molar density of fluid molecules in the coal pores per unit volume of undeformed porous medium.

In the coal pores, the molecules of fluid can be either in their bulk state inside the pore (with a molar density n_p^b per unit volume of undeformed medium) or adsorbed at the surface of the coal pore (with a molar density n_p^a per unit volume of undeformed medium), so that $n_p = n_p^b + n_p^a$. In contrast, in the cleats, all molecules of fluid are in a bulk state. We now introduce the Helmholtz free energy f_l of the fluids in a bulk state per unit volume of undeformed porous medium. By definition, we have:

$$f_l = n_c \mu_c - p_c \phi_c + n_p^b \mu_p - p_p \phi_p \quad [7.10]$$

Combining equations [7.9] and [7.10] and using the Gibbs–Duhem relations $n_c d\mu_c - \phi_c dp_c = 0$ and $n_p^b d\mu_p - \phi_p dp_p = 0$, we obtain the energy balance for the system made of the reservoir without its bulk fluids [COU 10, VAN 10]:

$$d(f - f_l) = \sigma d\varepsilon + s_{ij} de_{ij} + p_c d\phi_c + p_p d\phi_p + \mu_p dn_p^a \quad [7.11]$$

Identifying that the reservoir without its bulk fluids is in fact made of the solid skeleton and of the surface of the pores, we find that $f - f_l = f_{skel} + f_{surf}$, where f_{skel} is the Helmholtz free energy stored in the solid skeleton per unit volume of undeformed porous medium. Combining equation [7.11] with equation [7.3] eventually yields the energy balance for the solid skeleton [COU 10, VAN 10]:

$$df_{skel} = \sigma d\varepsilon + s_{ij}de_{ij} + p_c d\phi_c + p_p d\phi_p - \tilde{\sigma}^s ds_p \qquad [7.12]$$

For an isotropic medium under small deformations, for reasons of symmetry, the surface of the coal pores cannot depend on the deviatoric strain. Moreover, the surface s_p of the coal pores must only depend on the variables that define the state of the coal material, i.e. the coal porosity ϕ_p and the volumetric strain ε_p of the coal material: $s_p = s_p(\varepsilon_p, \phi_p)$. A macroscopic strain being the space average of its microscopic counterparts [COU 10], we have $\varepsilon = (1 - \phi_{c0})\, \varepsilon_p + \phi_c - \phi_{c0}$ (where ϕ_{c0} is the porosity associated to the cleat system in the state of reference), from what follows: $s_p = s_p(\varepsilon - \phi_c, \phi_p)$. Therefore, $ds_p = -c_{\Phi_c} d\varepsilon + c_{\Phi_c} d\phi_c + c_{\Phi_p} d\phi_p$, where c_{Φ_c} and c_{Φ_p} are, in first order, constant material parameters. The energy balance [7.12] can therefore be rewritten as:

$$df_{skel} = (\sigma + p_c^a)d\varepsilon + s_{ij}de_{ij} + (p_c - p_c^a)d\phi_c + (p_p - p_p^a)d\phi_p \qquad [7.13]$$

where $p_c^a = c_{\Phi_c}\tilde{\sigma}^s$ and $p_p^a = c_{\Phi_p}\tilde{\sigma}^s$.

Comparing energy balance [7.13] with energy balance [7.5], the constitutive equations for a reservoir in which adsorption occurs can readily be inferred from the constitutive equations [7.6] for a reservoir in which no adsorption occurs, by replacing in the latter σ with $\sigma + p_c^a$, p_c with $p_c - p_c^a$, and p_p with $p_p - p_p^a$. In the final step, imposing an identical pressure of the fluid in the cleats and in the coal pores (i.e. $p_c = p_p = p$), and noting $b_c + b_p = b$, we obtain the constitutive equations for our dual-porosity reservoir:

$$\begin{cases} d\sigma = Kd\varepsilon - bdp + b_c dp_c^a + b_p dp_p^a - dp_c^a \\ ds_{ij} = 2Gde_{ij} \\ d\phi_c = b_c d\varepsilon + (dp - dp_c^a)/N_c - (dp_c^a - dp_p^a)/N_{cp} \\ d\phi_p = b_p d\varepsilon + (dp - dp_p^a)/N_p - (dp_p^a - dp_c^a)/N_{cp} \end{cases} \qquad [7.14]$$

where the poroelastic parameters of the reservoir depend on the confining stress σ and on the pressure p of the fluid, and where the micromechanical relations [7.7] and [7.8] still hold. Therefore, surface adsorption modifies the poromechanical behavior of the reservoir through the introduction of a pre-pore pressure $p_c^a = c_{\Phi_c}\tilde{\sigma}^s$ which acts in the cleat system and of a pre-pore pressure $p_p^a = c_{\Phi_p} \tilde{\sigma}^s$ which acts in the coal pore system. How adsorption modifies the surface stress $\tilde{\sigma}^s$ is governed by Gibbs' law [7.2].

7.3. Simulations

Section 7.2 was dedicated to the derivation of the constitutive equations of a coal bed reservoir in which adsorption-induced swelling occurs. In order to implement the constitutive equations [7.14] in a solver, we need to know how the surface stress $\tilde{\sigma}^s$ depends on adsorption. This dependence is given by Gibbs law [7.2], which requires the adsorption isotherm $\Gamma(\mu)$ (or, equivalently, $\Gamma(p)$) as input. Therefore, as a pre-requisite to any reservoir simulation of a coal bed injected with carbon dioxide, we need to determine the adsorption isotherm of carbon dioxide on coal. In the following section we estimate such an adsorption isotherm with molecular simulations.

7.3.1. *Simulations at the molecular scale: adsorption of carbon dioxide on coal*

We performed Monte Carlo simulations of adsorption of carbon dioxide on an atomistic structure representative of a real coal [VAN 10]. The simulations were performed in the grand canonical ensemble, in which the molar chemical potential μ of the molecules of carbon dioxide, the volume V of the box of simulation, and the temperature T are held constant. The temperature T was fixed at 310 K, and several molar chemical potentials were used, in order to cover bulk fluid pressures ranging from 0 to 20 MPa.

The simulations of adsorption were performed on a rigid molecular structure called CS1000 [JAI 06a]. CS1000 is a cube with 2.5 nm-long edges. Its structure is that of a high density porous saccharose coke obtained by pyrolyzing pure saccharose at 1000°C in a nitrogen flow. CS1000 is amorphous and includes micropores in which only a few molecules of carbon dioxide can fit. The physical properties of CS1000 are relatively close to that of a real coal. Indeed, the porosity in the CS1000 sample, probed by a spherical particle with a 3Å diameter, is 14%, which is in good agreement with experimental measurements of the porosity of low rank or medium rank coals [MAR 09]. The CS1000 model has a density of 1,584 $kg{\cdot}m^{-3}$: such a value is in the high range of the helium densities of coal, which are usually between 1,250 kg m^{-3} and 1,600 kg m^{-3} [SEN 01]. CS1000 only partially takes into account the chemical heterogeneity of real coal: it contains carbon atoms and hydrogen atoms (as coal does), but contains no oxygen atoms, while real coal contains from about 5% to more than 20% (in mass) of oxygen atoms [VAS 96].

The point charges in the CS1000 sample were determined with the "partial equalization of orbital electronegativity" method [GAS 80].

Carbon dioxide was modeled with the EPM model [HAR 95], in which each molecule is made of three Lennard–Jones centers corresponding to the carbon atom and to the two oxygen atoms, with one point charge at each of the centers. The molecule is rigid and linear, with the carbon atom in between the oxygen atoms. The energy $U_{CO_2-CO_2}$ of interaction between two molecules of carbon dioxide is:

$$U_{CO_2-CO_2} = \sum_{a,b}\left[\frac{q_a q_b}{4\pi\varepsilon_0 r_{ab}} + 4\varepsilon_{ab}\left(\left(\frac{\sigma_{ab}}{r_{ab}}\right)^{12} - \left(\frac{\sigma_{ab}}{r_{ab}}\right)^{6}\right)\right] \quad [7.15]$$

where the sum is performed over the pairs (a,b) of carbon atoms or oxygen atoms, a holds for the first atom (atom of carbon or of oxygen) of the pair and b for the second atom (atom of carbon or of oxygen) of the pair, r_{ab} is the distance between the atoms, q_j is the point charge on atom j, ε_0 is the permittivity of vacuum, $\varepsilon_{CC} = 28.129\ k_B$, $\varepsilon_{OO} = 80.507\ k_B$, $\varepsilon_{CO} = 45.588\ k_B$, k_B is the Boltzmann constant, $\sigma_{CC} = 2.757$Å, $\sigma_{OO} = 3.033$Å, $\sigma_{CO} = 2.892$Å, $q_C = 0.6512\ e$, $q_O = -\ q_C/2$, e is the elementary charge, and the distance between the carbon and oxygen atoms is $L_{CO} = 1.149$Å.

The interaction of carbon dioxide with the atoms of the CS1000 sample includes van der Waals interaction, short range repulsion, and electrostatic interaction. This interaction was modeled with the following potential U_{CO_2-i} of interaction:

$$U_{CO_2-i} = \sum_{a}\left[\frac{q_a q_i}{4\pi\varepsilon_0 r_{ai}} + 4\varepsilon_{ai}\left(\left(\frac{\sigma_{ai}}{r_{ai}}\right)^{12} - \left(\frac{\sigma_{ai}}{r_{ai}}\right)^{6}\right)\right] \quad [7.16]$$

where the sum is performed over the atom a (atom of carbon or of oxygen) of the molecule of carbon dioxide, i stands for the atom (atom of carbon or of hydrogen) considered in the CS1000 sample, r_{ai} is the distance between the atoms, q_a is the point charge on atom a in the molecule of carbon dioxide, q_i is the point charge on atom i in the CS1000 sample, ε_{ai} and σ_{ai} are Lennard–Jones parameters obtained with the Lorentz–Berthelot rules by using $\varepsilon_C = 28\cdot k_B$ and $\sigma_C = 3.36$Å for the carbon atoms of CS1000, and $\varepsilon_H = 15.08\cdot k_B$ and $\sigma_H = 3.36$Å for the hydrogen atoms of CS1000 [JAI 06b].

At each chemical potential considered, the molecular simulation provided the total amount of molecules of fluid adsorbed in the sample. In order to use Gibbs' law [7.2], we needed to determine the adsorbed amount in excess of the bulk density per unit area of the surface of the coal pores. To do so, we needed to subtract the amount of fluid that would fill the pore volume if the fluid was at its bulk density from the total amount of fluid in the sample. We defined the pore volume as the maximal volume occupied by the molecules of carbon dioxide among all simulations, assuming that those molecules were at a liquid density. Considering a liquid density for carbon dioxide of 2.35×10^4 mol·m^{-3}, we estimated the porosity at 9%. Then, considering that each molecule of carbon dioxide covered a surface of 16.9 Å^2, we estimated the specific surface of the CS1000 sample at 2.19×10^8 m$^2 \cdot$ m^{-3}. Based on those estimates, we computed the excess adsorbed amount Γ per unit area of the pore surface. The results are displayed in Figure 7.2.

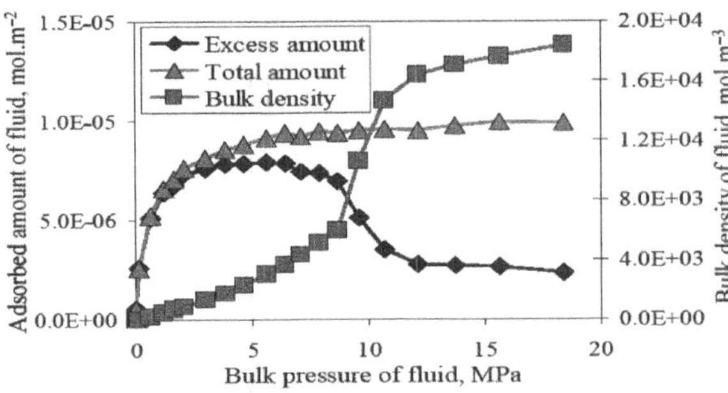

Figure 7.2. *Results of the molecular simulations of adsorption of carbon dioxide in the CS1000 sample. The "total amount" is the total amount of molecules of fluid in the sample. The "excess amount" is the amount of molecules of fluid adsorbed in the sample in excess of the bulk density of the fluid. Both amounts are expressed per unit area of the surface of the coal pores*

At fluid pressures lower than 7 MPa, the adsorption was characteristic of a monolayer site adsorption and was nearly insensitive to the estimate of porosity, since the bulk density of carbon dioxide was small in this range of pressure. At larger pressures, the total amount of fluid in the sample remained nearly constant, but the bulk density of the fluid significantly increased: as a consequence, the adsorbed amount in excess of the bulk density decreased significantly.

7.3.2. *Simulations at the scale of the reservoir*

Taking into account the adsorption isotherm estimated in section 7.3.1, we aim at performing reservoir simulations of an injection of carbon dioxide into a coal bed that contains only carbon dioxide.

As a first step, we only simulated a representative elementary volume of reservoir. The volume was constrained, meaning that no deformation was allowed. The pressure of the fluid inside the volume was homogeneous and was increased from 0 to 18 MPa.

The material was assumed to behave in a linear elastic manner, with the following material parameters: K_s = 2 GPa, K_m = 1.7 GPa, K = 0.67 GPa, ν = 0.25, $\phi_c (p = 1$ MPa) = 3%, $\phi_p (p = 1$ MPa) = 2%, $c_{\phi_c} = 0$ m^{-1}, c_{Φ_p} = 2.21 × 10 m^{-1}. The values for c_{ϕ_c} and c_{ϕ_p} were obtained by fitting experimental data of swelling of coal in presence of carbon dioxide obtained by Levine [LEV 96] and assuming that coal is a porous matrix with spherical pores [VAN 10]. The equation of state for carbon dioxide was obtained by molecular simulations with the EPM model and is displayed in Figure 7.2. The viscosity of carbon dioxide was η = 1.79 × 10^{-5} Pa·s. The adsorption isotherm of carbon dioxide on coal was the isotherm obtained by molecular simulations as discussed in section 7.3.1. The permeability k of the REV was obtained with a Kozeny–Carman-type equation based on the cleat porosity only: $k \propto \phi_c^3 / (1-\phi_c)^2$.

The equations described above were solved using the nonlinear finite element and volume software Bil. The results of the simulations are displayed in Figure 7.3. Figure 7.3(a) shows that, as the bulk pressure of the fluid increased, the surface stress which acts at the surface of the coal pores decreased. Such a decrease is consistent with Gibbs' law [7.2]. As a consequence of the decrease of the surface stress, the coal porosity increased with the bulk fluid pressure (see Figure 7.3(b)). Interestingly, Figure 7.3(b) also shows that the cleat porosity did not evolve monotonically with the bulk fluid pressure: at low pressure of the fluid (below 5 MPa), the cleat porosity decreased with the fluid pressure, while at high pressure of the fluid (above 5 MPa), the cleat porosity increased with the fluid pressure. Such a phenomenon is the result of two competing phenomena: on the one hand, the swelling of coal tends to close the cleats; on the other hand, the pressure of the fluid in the cleats tends to open the cleats. Likewise, the permeability of the REV did not evolve monotonically with the pressure of the fluid (see Figure 7.3(c)): at low pressure of the fluid, the adsorption-induced swelling of the coal material is the prevailing phenomenon, which closes the cleats and thus induces a decrease of permeability, while at higher pressure the fluid reopens the cleats, which translates into a permeability that again increases. Such a phenomenon is called the permeability rebound [SHI 04].

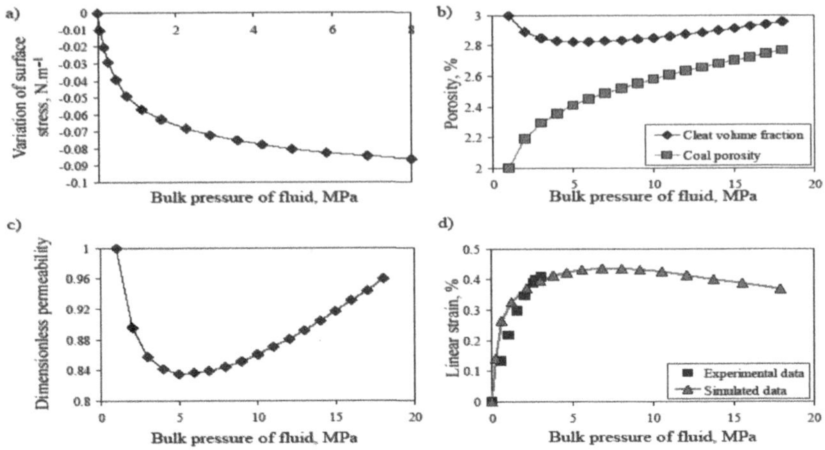

Figure 7.3. *Results of the simulations performed on an REV (representative elementary volume) and based on the adsorption isotherm estimated by molecular simulations: a) variation of the surface stress acting at the surface of the pores in a constrained REV, b) cleat porosity and coal porosity of a constrained REV, c) dimensionless permeability of a constrained REV, and d) swelling of an REV immersed in carbon dioxide (the experimental data are adapted from [LEV 96])*

We then simulated an REV immersed in carbon dioxide at a given pressure. We considered drained conditions and imposed $\sigma = -p$ as a boundary condition. The volumetric swelling obtained from the simulations is displayed in Figure 7.3(d), together with experimental data obtained by Levine [LEV 96]. The simulated results compare reasonably well with the experimental data. Up to a fluid pressure of 7 MPa, the sample swelled with an increasing fluid pressure, as a direct consequence of the adsorption of carbon dioxide in the coal porosity and of the consecutive release of surface stress. At fluid pressures greater than 7 MPa, the swelling of the sample decreased with an increasing fluid pressure, which is due to the compressibility of the sample.

We then performed reservoir simulations of an injection of carbon dioxide into a coal bed containing carbon dioxide only. The simulations were axisymmetric and one-dimensional with plane-strain conditions. The radius of the reservoir was 40 m, and the radius of the bore hole was 0.1 m. Before injection, the pressure of the carbon dioxide in the reservoir was equal to 1 MPa. At this pressure, the permeability in the reservoir was 10 mD everywhere. At time $t = 0$, the pressure of the carbon dioxide in the bore hole was increased immediately to 6 MPa. On the edge of the reservoir, we imposed no displacement and no flow.

Two different cases were simulated: in a first case, adsorption-induced swelling was explicitly taken into account by using the material parameters given above; in a second case, adsorption-induced swelling was prevented by setting $c_{\Phi_p} = 0\ \text{m}^{-1}$. The resulting rates of injection are displayed in Figure 7.4. We observed a significant decrease of the rate of injection when adsorption-induced swelling was taken into account: such a swelling led to a closure of the cleat system, which translated into a decrease in the injectivity of the reservoir.

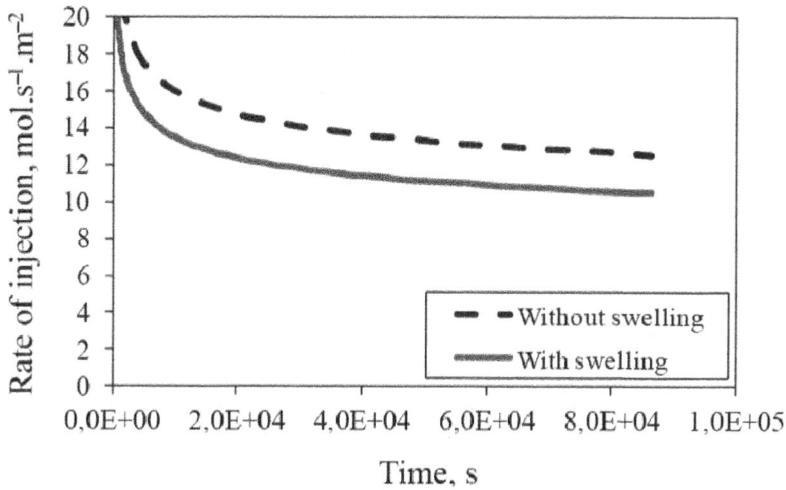

Figure 7.4. *Rate of injection of carbon dioxide into the reservoir when swelling occurs or without swelling*

7.3.3. *Discussion*

In this chapter, we combined the work at the scale of the molecule and at the scale of the reservoir.

Regarding the molecular simulations, we can wonder about their representativeness. Indeed, due to its size, CS1000 does not contain mesopores (i.e. pores that have a radius greater than 1 nm) whereas real coal does: therefore, the isotherms which we simulated disregard adsorption in mesopores. Such a disregard should not be an issue, since the adsorption-induced swelling of coal is mostly due to adsorption in micropores (i.e. pores that have a radius below 1 nm) [VAN 10]. Also, CS1000 does not contain oxygen atoms whereas real coal does. Therefore, the simulated isotherms could underestimate the adsorption observed in practice with carbon dioxide, since oxygen atoms carry a significant electrostatic charge.

The molecular simulations which we performed were Monte Carlo simulations, which provide information on equilibrium states but provide no information on the kinetics of a phenomenon. Moreover, in such simulations, in which molecules of fluid can be inserted anywhere in the box of simulation, all pores are accessible, independently of the connectivity of the pore network. Therefore, in the simulations, molecules of fluid could be adsorbed at the surface of pores to which they have no access in practice.

The poromechanical model which we derived for a coal bed reservoir is simple with respect to a real coal bed reservoir. Our model is isotropic, whereas a coal bed reservoir is clearly anisotropic because of the geometry of the cleat system. We only considered carbon dioxide in the reservoir, whereas the practical case of ECBM requires considering mixtures of at least carbon dioxide and methane [BRO 12a]. Moreover, the permeability model which we implemented is not the most relevant for fractured media. However, despite these simplifications, the model we propose enables us to capture the variations of injectivity (e.g. the permeability rebound) of a coal bed consecutive to an adsorption-induced swelling of the coal matrix.

Our model for the coal bed reservoir is based on poromechanical equations recently extended to surface effects [COU 10, VAN 10]. In such a model, we used Gibbs' theory and assumed that adsorption occurs by surface covering. But, we can wonder whether such a model is relevant for coal, in which the smallest pores can only welcome a few molecules of fluid: in such small pores, can we still use Gibbs' theory? Can we even define the surface of such small pores in an unequivocal manner? Deriving a more realistic model requires extending the realm of poromechanics to cases for which adsorption occurs by pore filling more than by surface covering, which is what we did recently [BRO 12b, NIK 12].

7.4. Conclusions

In this chapter, we studied the injection of carbon dioxide in a coal bed by combining work at the scale of the molecule and at the scale of the reservoir. We simulated adsorption isotherms of carbon dioxide on coal, derived the poromechanical constitutive equations of a coal bed reservoir, and implemented those constitutive equations together with the simulated adsorption isotherms in order to perform reservoir simulations.

We modeled the coal bed reservoir as a dual-porosity medium, and derived poromechanical equations that explicitly take into account surface adsorption. The model could be made more complex and realistic, for instance, by extending it to anisotropy or to multi-component fluids. Nevertheless, the derived constitutive equations enabled us to capture the evolutions of injectivity of the reservoir during the process of injection. The simulations illustrated how those variations of injectivity result from a competition between an adsorption-induced swelling (which tends to close the cleat system and thus to decrease the injectivity) and a pressure-induced opening of the cleats (which tends to increase the injectivity).

The poromechanical model for the reservoir needed the isotherms of adsorption of carbon dioxide on coal as an input. We obtained these isotherms of adsorption by performing Monte Carlo molecular simulations. Molecular simulations are an interesting alternative to experiment, since performing adsorption experiments is complex and tedious. In contrast, molecular simulations enable us to control well the inputs (temperature of the system, bulk pressure of the fluid, etc.) and to vary those inputs easily.

We used a multiscale approach: we aimed at describing and simulating a behavior at the macroscopic scale relevant to the engineer, and for that purpose also used molecular simulations in order to obtain a piece of information at the molecular scale. Such a multiscale approach, here exemplified on an injection of carbon dioxide in a coal bed, is relevant to any porous medium in which adsorption causes a strain.

7.5 Bibliography

[BRO 12a] BROCHARD L., VANDAMME M., PELLENQ R.J.-M., FEN-CHONG T., "Adsorption-induced deformation of microporous materials: coal swelling induced by CO_2-CH_4 competitive adsorption", *Langmuir*, vol. 28, pp. 2659–2670, 2012.

[BRO 12b] BROCHARD L., VANDAMME M., PELLENQ R.J.-M., "Poromechanics of microporous media", *Journal of the Mechanics and Physics of Solids*, vol. 60, pp. 606–622, 2012.

[CHE 10] CHEN Z., LIU J., ELSWORTH D., CONNELL L.D., PAN Z., "Impact of CO_2 injection and differential deformation on CO_2 injectivity under in-situ stress conditions", *International Journal of Coal Geology*, vol. 81, pp. 97–108, 2010.

[COU 04] COUSSY O., *Poromechanics*, John Wiley & Sons Ltd, Chichester, 2004.

[COU 10] COUSSY O., *Mechanics and Physics of Porous Solids*, John Wiley & Sons Ltd, Chichester, 2010.

[GAS 80] GASTEIGER J., MARSILI M., "Iterative partial equalization of orbital electronegativity—a rapid access to atomic charges", *Tetrahedron*, vol. 36, pp. 3219–3228, 1980.

[GIB 28] GIBBS J.W., *The Collected Works of J. Willard Gibbs*, Green and Co, New York, Longmans, 1928.

[HAR 95] HARRIS J.G., YUNG K.H., "Carbon dioxide's liquid-vapor coexistence curve and critical properties as predicted by a simple molecular model", *Journal of Physical Chemistry*, vol. 99, pp. 12021–12024, 1995.

[JAI 06a] JAIN S.K., PELLENQ R.J.-M., PIKUNIC J.P., GUBBINS K.E., "Molecular modeling of porous carbons using the hybrid reverse Monte Carlo method", *Langmuir*, vol. 22, pp. 9942–9948, 2006.

[JAI 06b] JAIN S.K., GUBBINS K.E., PELLENQ R.J.-M., PIKUNIC J.P., "Molecular modeling and adsorption properties of porous carbons", *Carbon*, vol. 44, pp. 2445–2451, 2006.

[LAP 06] DE LAPLACE P.S., *Traité de Mécanique Céleste*, Gauthier-Villars, Paris, 1806.

[LEV 96] LEVINE J.R., *Model Study of the Influence of Matrix Shrinkage on Absolute Permeability of Coal Bed Reservoirs*, Geological Society Special Publications, London, vol. 109, pp. 197–212, 1996.

[MAR 09] MARES T.E., RADLINSKI A.P., MOORE T.A., COOKSON D., THIYAGARAJAN P., ILAVSKY J., KLEPP J., "Assessing the potential for CO_2 adsorption in a subbituminous coal, Huntly Coalfield, New Zealand, using small angle scattering techniques", *International Journal of Coal Geology*, vol. 77, pp. 54–68, 2009.

[NIK 12] NIKOOSOKHAN S., VANDAMME M., DANGLA P., "A poromechanical model for coal seams injected with carbon dioxide: from an isotherm of adsorption to a swelling of the reservoir", *Oil & Gas Science and Technology*, accepted, 2012.

[REE 03] REEVES S., TAILLEFERT A., PEKOT L., CLARKSON C., The Allison unit CO_2 – ECBM pilot: a reservoir modeling study, Topical Report, U.S. Department of Energy, DE-FC26-0NT40924, 2003.

[SEN 01] SENEL A.G., GURUZ A.G., YUCEL H., KANDAS A.W., SAROFIM A.F., "Characterization of pore structure of turkish coals", *Energy & Fuels*, vol. 15, pp 331–338, 2001.

[SHI 04] SHI J.Q., DURUCAN S., "Drawdown induced changes in permeability of coal beds: a new interpretation of the reservoir response to primary recovery", *Transport in Porous Media*, vol. 56, pp. 1–16, 2004.

[SHU 50] SHUTTLEWORTH R., "The surface tension of solids", *Proceedings of the Physical Society A*, vol. 63, pp. 444–457, 1950.

[VAN 10] VANDAMME M., BROCHARD L., LECAMPION B., COUSSY O., "Adsorption and strain: the CO_2-induced swelling of coal", *Journal of the Mechanics and Physics of Solids*, vol. 58, pp. 1489–1505, 2010.

[VAS 96] VASSILEV S.V., KITANO K., VASSILEVA C.G., "Some relationships between coal rank and chemical and mineral composition", *Fuel*, vol. 75, pp. 1537–1542, 1996.

[WHI 05] WHITE C.M., SMITH D.H., JONES K.L., GOODMAN A.L., JIKICH S.A., LACOUNT R.B., DUBOSE S.B., OZDEMIR E., MORSI B.I., SCHROEDER K.T., "Sequestration of carbon dioxide in coal with enhanced coal bed methane recovery – a review", *Energy & Fuels*, vol. 19, pp. 659–724, 2005.

[WU 10] WU Y., LIU J., ELSWORTH D., CHEN Z., CONNELL L., PAN Z., "Dual poroelastic response of a coal seam to CO_2 injection", *International Journal of Greenhouse Gas Control*, vol. 4, pp. 668–678, 2010.

[WU 11] WU Y., LIU J., CHEN Z., ELSWORTH D., PONE D., "A dual poroelastic model for CO_2-enhanced coal bed methane recovery", *International Journal of Coal Geology*, vol 86, pp. 177–189, 2011.

[YOU 05] YOUNG T., "An essay on the cohesion of fluids", *Philosophical Transactions of the Royal Society of London*, vol. 95, pp. 65–87, 1805.

[ZHA 08] ZHANG H., LIU J., ELSWORTH D., "How sorption-induced matrix deformation affects gas flow in coal seams: a new FE model", *International Journal of Rock Mechanics and Mining Sciences*, vol. 45, pp. 1226–1236, 2008.

PART 3

Aging and Integrity

Chapter 8

Modeling by Homogenization of the Long-Term Rock Dissolution and Geomechanical Effects

8.1. Introduction

The geological storage of CO_2 coming from large industrial facilities has been studied in detail for several years as a solution against the greenhouse effect and climate change in a number of countries. It is considered as a complementary solution together with the research into non-pollutant and/or renewable energy sources. One of the available options is the injection of the supercritical CO_2 in the saline aquifers. The idea of geological storage resides in various mechanisms of trapping coming into play progressively over time. In the short term, the migration of CO_2 is principally blocked by the impervious rock lying on the top of the aquifer. With the passage of time, other mechanisms appear, in particular of geochemical nature, like the dissolution of CO_2 in pore water, and finally the chemical reactions resulting in dissolution and/or precipitation of rock minerals. These reactions can generate important and irreversible modifications of the hydrodynamical as well as geomechanical properties of the reservoir.

The long-term safety of the CO_2 storage sites requires the understanding of the physicochemical processes related to the injection of the CO_2 in the geological medium and their consequences, especially for the stability of the reservoir. These

---

Chapter written by Jolanta LEWANDOWSKA.

processes are still not completely known, in particular, because of the complexity of natural systems (very often heterogeneous and multiscale), and also because of multiphysical couplings engendered by the CO_2 injection.

The experimental studies of CO_2 injection into rock samples, conducted in the laboratory conditions, show different behavior depending on the injection rate, the composition of the injected fluids, and the nature of the rock (e.g. [LUQ 09, GOU 10, EGE 05, IZG 08]). The investigations focused on the long-term behavior assume that the rock is saturated with an acidified aqueous CO_2 solution, and homogeneous chemical alteration occurs. The results of experimental studies conducted in such conditions show global porosity increase homogeneously distributed over the sample length (e.g. [NGU 11, BER 09]). Moreover, it was observed that chemical alteration induces mechanical weakening of the rocks with a decrease of both stiffness and shear strength. The triaxial testing in drained conditions on the altered carbonate samples showed the decrease of the elastic bulk modulus and the shear modulus (e.g. [NGU 11, BER 09]). Although this behavior has been reported in the literature, very few contributions concerning the modeling of long-term geomechanical effects related to CO_2 storage are proposed. Of special interest in this case are the micromechanical methods because they provide suitable mathematical framework, capable to produce multiscale and multiphysics models of high predictive capacity.

8.2. Microstructure and modeling by homogenization

Let us assume that the microstructure of the porous rock is known; Figure 8.1. We consider a porous medium with two phases: solid skeleton and (saline) water saturating the pores. Both phases are assumed to be connected. The water contains the dissolved CO_2. The porous medium is periodic and the periodic volume is representative from the point of view of both the microstructure and the phenomena taking place at the microscopic scale. We assume that the conditions of scale separation are satisfied, which means that the size of the period (l) is very small with respect to the dimension of the aquifer (L). Mathematically, this condition is written in the form:

$$\varepsilon = \frac{\ell}{L} << 1 \qquad [8.1]$$

where ε is the small parameter of homogenization.

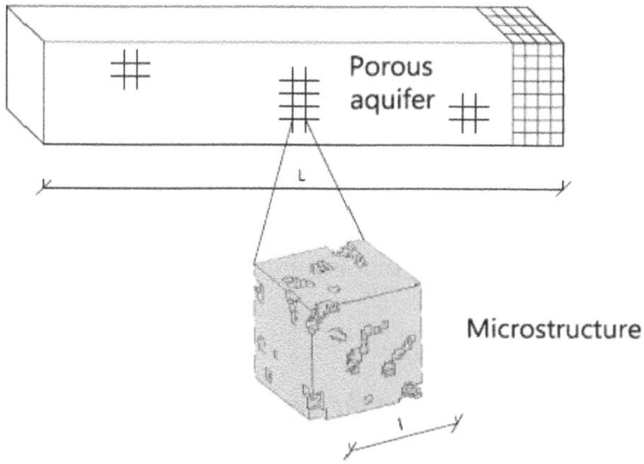

Figure 8.1. *The porous aquifer with the microstructure*

The following phenomena will be taken into account:

– the water flow through the pores (named hydraulic "H-problem");

– the solid deformation (named mechanical "M-problem");

– the transport of dissolved CO_2 (named transport "T-problem");

– the chemical dissolution of solid (named chemical "C-problem").

The aim of the analysis is to develop a macroscopic model of the behavior of the system by using the method of homogenization of periodic structures ([SAN 80, BEN 78]). The analysis will be presented in two stages. In the first stage (section 8.3), the homogenization of the problem of water flow, solid deformation, and CO_2 transport (the "H-M-T problem") will be analyzed. We will use the principal results of homogenization of the "H-M problem" and the "T problem" that were previously analyzed as uncoupled problems ([AUR 77, AUR 96]). In the second stage (section 8.4), the chemo-mechanical coupling will be considered (the "C-M problem"). At this stage it will be assumed that the whole aquifer is subjected to homogeneous solid dissolution caused by chemical reaction between solid components and acid pore water. We will focus on the modification of the microstructure of the porous aquifer and progressive decrease of the mechanical parameters. In section 8.5 numerical computations of the chemo-mechanical coupling and the chemical degradation of the rock mechanical parameters will be presented.

8.3. Homogenization of the H-M-T problem

To model the coupled H-M-T behavior at the macroscopic scale, the differential equations are first formulated at the microscopic (pore) scale. This formulation concerns the equilibrium equation for stresses in the solid phase, the momentum equation for fluid flow, the mass balance equation of concentration of the dissolved CO_2, and the appropriate interface conditions. Then, the classical homogenization process by using the asymptotic developments method [SAN 80, BEN 78] is carried out.

8.3.1. *Formulation of the problem at the microscopic scale*

The assumptions are as follows:

– The porous medium is saturated by viscous incompressible fluid (e.g. saline water).

– The fluid is saturated by the dissolved CO_2.

– The fluid flow is slow and laminar.

– The transport of the CO_2 dissolved in fluid is governed by convection and diffusion mechanism.

– Small displacements and deformations of the solid.

– Linear elastic constitutive law for solid material.

– Quasi-static deformation process.

The following set of differential equations can be written in the domain of the period (Figure 8.2):

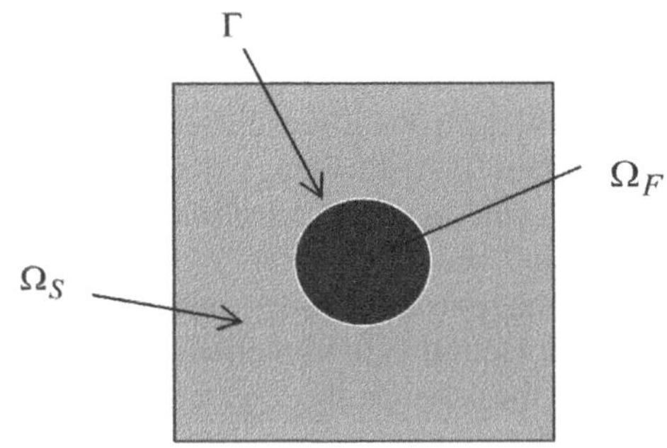

Figure 8.2. *The period of the porous medium*

– Stress equilibrium of the solid phase:

$$\frac{\partial \sigma_{Sij}}{\partial X_j} = 0 \; in \; \Omega_S \qquad [8.2]$$

$$\sigma_{Sij} = a_{ijkl}\, e_{kl}(\boldsymbol{u}) \qquad [8.3]$$

with $e_{kl} = 1/2(u_{k,l} + u_{l,k})$.

– Stress equilibrium and incompressibility condition of the fluid phase:

$$\frac{\partial \sigma_{Fij}}{\partial X_j} = 0 \; in \; \Omega_F \qquad [8.4]$$

$$\sigma_{Fij} = -p\delta_{ij} + \mu_F (v_{i,j} + v_{j,i}) \; in \; \Omega_F \qquad [8.5]$$

$$\frac{\partial v_i}{\partial X_i} = 0 \; in \; \Omega_F \qquad [8.6]$$

– Mass balance equation of the CO_2 dissolved in the fluid phase:

$$\frac{\partial c}{\partial t} + \frac{\partial}{\partial X_i}\left(-D_{ij} \frac{\partial c}{\partial X_j} + v_i c \right) = 0 \; in \; \Omega_F \qquad [8.7]$$

– Continuity conditions at the solid–fluid interface:

$$\sigma_{Sij} N_j - \sigma_{Fij} N_j = 0 \; in \; \Gamma \qquad [8.8]$$

$$\dot{u}_i - v_i = 0 \; on \; \Gamma \qquad [8.9]$$

– Zero-flux condition at the solid–fluid interface:

$$\left(-D_{ij} \frac{\partial c}{\partial X_j} \right) N_i = 0 \; on \; \Gamma \qquad [8.10]$$

where $\boldsymbol{\sigma}_S$ is the stress tensor in the solid, $\boldsymbol{\sigma}_F$ is the stress tensor in the fluid, $\boldsymbol{a}$ is the microscopic rigidity tensor of the solid, $\boldsymbol{e}$ is the small deformation tensor in the solid, $\boldsymbol{u}$ is the displacement vector in the solid, p is the water pressure, μ_F is the viscosity of the fluid, c is the concentration of the CO_2 dissolved in water, $\boldsymbol{D}$ is the diffusion tensor of the CO_2 in water, v is the water flow velocity, and $\mathbf{N}$ is the unit normal outward vector.

Before starting the homogenization, the descriptions [8.2]–[8.10] have to be rewritten using the non-dimensional variables. Here we adopt the so-called "microscopic point of view" [AUR 91], which means that the length scale used in the normalization will be the microscopic dimension l. For the physical (dimensional) space variable $\boldsymbol{X}$, two non-dimensional space variables are defined:

$$x = \frac{\boldsymbol{X}}{L} \text{ and } y = \frac{\boldsymbol{X}}{\ell}$$

The homogenization problem will be analyzed at two scales $\boldsymbol{x}$ and $\boldsymbol{y}$, simultaneously.

The introduction of the non-dimensional variables is related to the evaluation of the order of magnitude of several quantities appearing in the microscopic model [8.2]–[8.10]. As a unit of measure in this evaluation, the powers of the small parameter ε are used. The following estimations are supposed to be valid:

– The Reynolds number is: $\mathrm{Re}_l = \dfrac{\rho v \ell}{\mu_F} = O(\varepsilon^1)$ where ρ is the volumetric density of the fluid.

– The macroscopic gradient of water pressure is of the same order of magnitude as the microscopic viscous term: $p/L = O(\mu_F\, v / \ell^2)$.

– The local Péclet number is such that there is no dispersion: $Pe_l = \dfrac{v\ell}{D} = O(\varepsilon^1)$.

– The timescale t_C corresponds to the mass transport by diffusion at the macroscopic scale L, thus the associated non-dimensional number is $P_l = \dfrac{\ell^2}{D t_C} = O(\varepsilon^2)$.

– The displacements and stresses in solid and in fluid are of the same orders of magnitude: $u_{Si} = O(u_{Fi})$ and $\sigma_{Sij} = O(\sigma_{Fij})$.

More details concerning the evaluations of the orders of magnitude can be found in [AUR 91, AUR 77, AUR 96]. Finally, the non-dimensional description takes the following form:

$$\frac{\partial}{\partial y_j}(a_{ijkl}\, e_{ykl}(\boldsymbol{u})) = 0 \ in\ \Omega_S \qquad [8.11]$$

$$\varepsilon \mu_F \frac{\partial^2 v_i}{\partial y_j \partial y_j} - \frac{\partial p}{\partial y_i} = 0 \ in \ \Omega_F \quad [8.12]$$

$$\frac{\partial v_i}{\partial y_i} = 0 \ in \ \Omega_F \quad [8.13]$$

$$\varepsilon^2 \frac{\partial c}{\partial t} + \frac{\partial}{\partial y_i}\left(-D_{ij} \frac{\partial c}{\partial y_j} + v_i c \right) = 0 \ in \ \Omega_F \quad [8.14]$$

The interface conditions are rewritten in the non-dimensional form as:

$$(a_{ijkl} \, e_{kl}(\boldsymbol{u}) + p\delta_{ij} - \varepsilon \mu_F (v_{i,j} + v_{j,i}))N_j = 0 \ on \ \Gamma \quad [8.15]$$

$$\dot{u}_i - v_i = 0 \ on \ \Gamma \quad [8.16]$$

$$\left(-D_{ij} \frac{\partial c}{\partial X_j} \right) N_i = 0 \ on \ \Gamma \quad [8.17]$$

The interface conditions [8.15]–[8.17] are completed by the periodicity conditions on the external boundaries of the period.

For convenience we keep the same notations for the non-dimensional variables in [8.11]–[8.17] as for the dimensional variables in [8.2]–[8.10]. Further, we introduce the relative velocity $w_i = v_i - \dot{u}_i$ which is the velocity of the fluid with respect to the solid. Using the new definition of fluid velocity the problem can be rewritten in the form of:

$$\frac{\partial}{\partial y_j}(a_{ijkl} \, e_{ykl}(\boldsymbol{u})) = 0 \ in \ \Omega_S \quad [8.18]$$

$$\varepsilon \mu_F \frac{\partial^2 w_i}{\partial y_j \partial y_j} - \frac{\partial p}{\partial y_i} = 0 \ in \ \Omega_F \quad [8.19]$$

$$\frac{\partial w_i}{\partial y_i} = 0 \ in \ \Omega_F \quad [8.20]$$

$$\varepsilon^2 \frac{\partial c}{\partial t} + \frac{\partial}{\partial y_i}\left(-D_{ij} \frac{\partial c}{\partial y_j} + w_i c \right) = 0 \ in \ \Omega_F \quad [8.21]$$

where the boundary conditions are:

$$(a_{ijkl}\, e_{kl}(\boldsymbol{u}) + p\delta_{ij} - \varepsilon\,\mu_F(w_{i,j} + w_{j,i}))N_j = 0 \; on \; \Gamma \tag{8.22}$$

$$w_i = 0 \; on \; \Gamma \tag{8.23}$$

$$\left(-D_{ij}\frac{\partial c}{\partial X_j}\right)N_i = 0 \; on \; \Gamma \tag{8.24}$$

and the periodicity conditions of all unknowns hold.

8.3.2. *Asymptotic developments method*

In the homogenization process, the unknowns (i.e. solid displacement, fluid velocity and fluid pressure) are presented in the form of asymptotic developments [SAN 80, BEN 78].

$$u_i(x,y,t) = u_i^{(0)}(x,y,t) + \varepsilon\, u_i^{(1)}(x,y,t) + \varepsilon^2\, u_i^{(2)}(x,y,t) + \ldots \tag{8.25}$$

$$p(x,y,t) = p^{(0)}(x,y,t) + \varepsilon\, p^{(1)}(x,y,t) + \varepsilon^2\, p^{(2)}(x,y,t) + \ldots \tag{8.26}$$

$$w_i(x,y,t) = w_i^{(0)}(x,y,t) + \varepsilon\, w_i^{(1)}(x,y,t) + \varepsilon^2\, w_i^{(2)}(x,y,t) + \ldots \tag{8.27}$$

Each term in the asymptotic development is assumed periodic, which means that it takes the same values in the corresponding points at the opposite walls of the period. Because of the existence of two separate scales, the derivation operator in [8.18]–[8.24] is transformed into $\partial/\partial y \rightarrow \partial/\partial y + \varepsilon\,\partial/\partial x$. Note that we have:

$$e_{xij}(\boldsymbol{u}) = \frac{1}{2}\left(\frac{\partial u_i}{\partial x_j} + \frac{\partial u_j}{\partial x_i}\right) \text{ and } e_{yij}(\boldsymbol{u}) = \frac{1}{2}\left(\frac{\partial u_i}{\partial y_j} + \frac{\partial u_j}{\partial y_i}\right).$$

In the process of homogenization, the asymptotic developments [8.25]–[8.27] are introduced in the microscopic description [8.18]–[8.24], while simultaneously applying the two-scale derivation operator. After identification of all terms staying by the same powers of ε, we obtain a series of successive order problems to be solved in the period. Typically, the first problem gives the solution for the term of order zero (for example in case of displacement: the form of $\boldsymbol{u}^{(0)}$ ($\boldsymbol{x}, \boldsymbol{y}, \boldsymbol{t}$)). From the second problem, we obtain the solution for the term of order one (e.g. in case of displacement: the form of $\boldsymbol{u}^{(1)}$ ($\boldsymbol{x}, \boldsymbol{y}, \boldsymbol{t}$))

that means the first corrector. Finally, the third problem is subjected to volume averaging to obtain the macroscopic description (model).

In section 8.3.3, the solutions of the homogenization process are presented. For more details concerning the homogenization of the problem [8.18]–[8.20] with the conditions [8.22] and [8.23], refer to [AUR 77]. For problems [8.21] and [8.24], refer to [AUR 96].

8.3.3. *Solutions*

8.3.3.1. *The zero-order solutions*

The analysis starts with the solution for the leading order of the main variables: $p^{(0)}$, $u^{(0)}$ and $c^{(0)}$.

The first problem is for pressure $p^{(0)}$ and comes from [8.19]:

$$\frac{\partial p^{(0)}}{\partial y_i} = 0 \ in \ \Omega_F \qquad [8.28]$$

where $p^{(0)}$ is periodic. The solution to this problem is $p^{(0)}(\boldsymbol{x},\boldsymbol{y},t) = p^{(0)}(\boldsymbol{x},t)$.

The second problem defines the displacement $u^{(0)}$ and comes from [8.18] and [8.22]:

$$\frac{\partial}{\partial y_j}(a_{ijkl}\, e_{ykl}(\boldsymbol{u}^{(0)})) = 0 \ in \ \Omega_S \qquad [8.29]$$

$$(a_{ijkl}\, e_{kl}(\boldsymbol{u}^{(0)}))N_j = 0 \ on \Gamma \qquad [8.30]$$

where $\boldsymbol{u}^{(0)}$ is periodic. The solution is $u^{(0)}(\boldsymbol{x},\boldsymbol{y},t) = u^{(0)}(\boldsymbol{x},t)$.

The third problem is for $c^{(0)}$ and is given by [8.21] and [8.24]:

$$\frac{\partial}{\partial y_i}\left(-D_{ij}\frac{\partial c^{(0)}}{\partial y_j}\right) = 0 \ in \ \Omega_F \qquad [8.31]$$

$$\left(-D_{ij}\frac{\partial c^{(0)}}{\partial X_j}\right)N_i = 0 \ on \ \Gamma \tag{8.32}$$

where $c^{(0)}$ is periodic. The solution is $c^{(0)}(x, y, t) = c^{(0)}(x, t)$.

We can see that for each physical phenomenon, we defined a macroscopic variable that does not depend on the microscopic space variable y.

The analysis is continued at the next order for the problems [8.19], [8.20] and [8.23], to obtain a solution for the relative flow velocity $w^{(0)}(x, y)$:

$$\mu_F \frac{\partial^2 w_i^{(0)}}{\partial y_j \partial y_j} - \frac{\partial p^{(1)}}{\partial y_i} - \frac{\partial p^{(0)}}{\partial y_i} = 0 \ in \ \Omega_F \tag{8.33}$$

$$\frac{\partial w_i^{(0)}}{\partial y_i} = 0 \ in \ \Omega_F \tag{8.34}$$

$$w_i^{(0)} = 0 \ on \ \Gamma \tag{8.35}$$

where $p^{(1)}$ and $w_i^{(0)}$ are periodic. The solution to the problems [8.33]–[8.35] is:

$$w_i^{(0)} = -\mathrm{k}_{ij}\frac{\partial p^{(0)}}{\partial x_j} \tag{8.36}$$

If we substitute $w_i^{(0)}$, according to $w_i = v_i - \dot{u}_i$, and calculate the volume average of the relative velocity, we obtain the Darcy law in the form [AUR 77]:

$$< w_i^{(0)} > = < v_i^{(0)} > - n\dot{u}_i = -K_{ij}\frac{\partial p^{(0)}}{\partial x_j} \tag{8.37}$$

where $K_{ij} = < k_{ij} >$ is the macroscopic permeability tensor. It can be computed from the local boundary value problem that arises, after the substitution of [8.36] into [8.33]–[8.35] [AUR 05, AUR 77].

8.3.3.2. *The first-order solutions*

The next problem concerning the solid behavior comes from [8.18] and [8.22] and defines $u_i^{(1)}$ as:

$$\frac{\partial}{\partial y_j}(a_{ijkl}(e_{ykl}(\boldsymbol{u}^{(1)})+e_{xkl}(\boldsymbol{u}^{(0)})))=0 \ in \ \Omega_S \qquad [8.38]$$

$$(a_{ijkl}(e_{ykl}(\boldsymbol{u}^{(1)})+e_{xkl}(\boldsymbol{u}^{(0)})))N_j=-p^{(0)}N_i \ on \ \Gamma \qquad [8.39]$$

where $u_i^{(1)}$ is periodic. The solution is a linear function as follows:

$$u_i^{(1)}=\xi_i^{kl}e_{xkl}(\boldsymbol{u}^{(0)})-\eta_i\, p^{(0)}+\overline{u}_i^{(1)}(\boldsymbol{x},t) \qquad [8.40]$$

where $\overline{u}_i^{(1)}(\boldsymbol{x},t)$ is an arbitrary vector. The characteristic (microscopic) functions $\boldsymbol{\xi}$ and $\boldsymbol{\eta}$ are a third-order tensor and a vector, respectively. They can be computed from the local boundary problem if we substitute [8.40] into [8.38] and [8.39] [AUR 77, ENE 84].

The next order approximation of the concentration $c^{(1)}$ is the solution of a problem obtained from [8.21] and [8.24]:

$$-\frac{\partial}{\partial y_j}\left(D_{ij}\left(\frac{\partial c^{(1)}}{\partial y_j}+\frac{\partial c^{(0)}}{\partial x_j}\right)\right)=0 \ in \ \Omega_F \qquad [8.41]$$

$$-\left(D_{ij}\left(\frac{\partial c^{(1)}}{\partial y_j}+\frac{\partial c^{(0)}}{\partial x_j}\right)\right)N_j=0 \ in \ \Omega_F \qquad [8.42]$$

where $c^{(1)}$ is periodic. In [8.41] and [8.42] we take into account $c^{(0)}(\boldsymbol{x},t)$. The solution is a linear function of the macroscopic gradient of concentration:

$$c^{(1)}=\zeta_i\frac{\partial c^{(0)}}{\partial x_i}+\overline{c}^{(1)}(\boldsymbol{x},t) \qquad [8.43]$$

where $\overline{c}^{(1)}(\boldsymbol{x},t)$ is an arbitrary function. The characteristic vectorial function $\boldsymbol{\zeta}$ can be calculated, if we introduce the solution [8.43] into problem [8.41] and [8.42]. For details refer [AUR 96].

8.3.3.3. *The macroscopic equations*

From [8.18], [8.19] and [8.22] at the next order we obtain the following set of equations:

$$\frac{\partial \sigma_{Sij}^{(1)}}{\partial y_j} + \frac{\partial \sigma_{Sij}^{(0)}}{\partial x_j} = 0 \ in \ \Omega_S \qquad [8.44]$$

$$\frac{\partial \sigma_{Fij}^{(1)}}{\partial y_j} - \frac{\partial p^{(0)}}{\partial x_i} = 0 \ in \ \Omega_F \qquad [8.45]$$

$$\sigma_{Sij}^{(1)} N_j = \sigma_{Fij}^{(1)} N_j \ on \ \Gamma \qquad [8.46]$$

with the periodicity conditions.

To obtain the macroscopic equations, equations [8.44] and [8.45] are integrated over the respective domains, and summed. Then, the boundary condition [8.46] is applied, that leads to the elimination of the derivatives with respect to the microscopic variable y. Finally, the macroscopic equilibrium equation is derived [AUR 77]:

$$< \frac{\partial \sigma_{ij}^{T(0)}}{\partial x_j} > = 0 \qquad [8.47]$$

where $\sigma^{T(0)}$ is the total stress tensor that is written as $\sigma_S^{(0)}$ and $\sigma_F^{(0)}$ within the respective domains Ω_S and Ω_F. After several transformations consisting mainly of the introduction of the solution for $u_i^{(1)}$ in the expression for stresses in the solid domain, we can write the total stress tensor in the form [AUR 77]:

$$\sigma_{ij}^{T(0)} = C_{ijkl} \ e_{xkl}(u^{(0)}) - \alpha_{ij} \ p^{(0)} \qquad [8.48]$$

where:

$$\sigma_{Sij}^{(0)} = C_{ijkl} \ e_{xkl}(u^{(0)}) \qquad [8.49]$$

and the macroscopic elasticity tensors C_{ijkl} and α_{ij} mean the rigidity tensor and the hydro-mechanical coupling elasticity tensor, respectively [AUR 77]:

$$C_{ijkl} = < a_{ijkl} + a_{ijmn} \ e_{ymn}(\xi^{kl}) > \qquad [8.50]$$

$$\alpha_{ij} = n I_{ij} + < a_{ijmn} \ e_{ymn}(\eta) > \qquad [8.51]$$

Both tensors can be computed for a given microstructure of the period, after the solution of the local boundary problems for ξ and η.

The second macroscopic equation is obtained from equation [8.20]. At the order ε it gives:

$$\frac{\partial w_i^{(0)}}{\partial x_i}+\frac{\partial w_i^{(1)}}{\partial y_i}=0 \ in\ \Omega_F \qquad [8.52]$$

Next, equation [8.52] is integrated within the domain Ω_F. We take into account the continuity of displacement on Γ, equation [8.23], and the previously obtained solution for $\boldsymbol{u}^{(0)}$. Finally, the following macroscopic equation is obtained [AUR 77]:

$$\frac{\partial(< v_i^{(0)} > - n\dot{u}_i)}{\partial x_i}=-\gamma_{ij}\,\dot{e}_{xij}(\boldsymbol{u}^{(0)})-\beta\,\dot{p}^{(0)} \qquad [8.53]$$

where γ and β are the two new coupling tensor and vector that are defined as follows:

$$\gamma_{ij}=n\,I_{ij}-<\xi_{p,p}^{ij}> \qquad [8.54]$$

$$\beta=<\eta_{p,p}> \qquad [8.55]$$

It can be shown that $\gamma=\boldsymbol{\alpha}$ [AUR 77].

Concerning the transport of the dissolved CO_2 from [8.21] and [8.24] we have:

$$\frac{\partial c^{(0)}}{\partial t}-\frac{\partial}{\partial x_i}\left(D_{ij}\left(\frac{\partial c^{(0)}}{\partial x_j}+\frac{\partial c^{(1)}}{\partial y_j}\right)\right)-\frac{\partial}{\partial y_i}\left(D_{ij}\left(\frac{\partial c^{(1)}}{\partial x_j}+\frac{\partial c^{(2)}}{\partial y_j}\right)\right)+$$
$$\frac{\partial}{\partial x_i}(w_i^{(0)}c^{(0)})+\frac{\partial}{\partial y_i}(w_i^{(1)}c^{(0)}+w_i^{(0)}c^{(1)})=0 \ in\ \Omega_F \qquad [8.56]$$

$$-\mathrm{N}_i\left(\frac{\partial c^{(1)}}{\partial x_j}+\frac{\partial c^{(2)}}{\partial y_j}\right)=0 \ on\ \Gamma \qquad [8.57]$$

To obtain the macroscopic equation, equation [8.56] is integrated over the volume Ω_F. Next, the interface condition [8.57], the periodicity equation [8.23] and the solution for $c^{(0)}$ are applied. After introduction of the solution [8.43], the final form of the macroscopic diffusion–convection equation is obtained [AUR 96]:

$$n\frac{\partial c^{(0)}}{\partial t}-\frac{\partial}{\partial x_i}\left(D_{ij}^{*}\frac{\partial c^{(0)}}{\partial x_j}\right)+\frac{\partial}{\partial x_i}(c^{(0)}< w_i^{(0)} >)=0 \qquad [8.58]$$

where the macroscopic diffusion tensor D^*_{ij} is defined as:

$$D^*_{ij} = \frac{1}{|\Omega|} \int_{\Omega_F} D_{ik} \left(I_{kj} + \frac{\partial \zeta_j}{\partial y_k} \right) d\Omega \qquad [8.59]$$

and $< w_i^{(0)} >$ is the Darcy velocity.

8.3.4. *Summary of the macroscopic "H-M-T model"*

Let us summarize the results of homogenization performed in section 8.3. The macroscopic model of coupled fluid flow, CO_2 transport and solid deformation ("H-M-T model") is defined by a set of macroscopic equations: [8.47], [8.48], [8.53], [8.37] and [8.58]. The model shows two kinds of couplings that can be called "weak" and "strong" coupling. The "weak" coupling can be observed between the fluid flow and the CO_2 transport, through the water velocity appearing in [8.58]. The "strong" coupling is the hydro-mechanical coupling that can be seen in [8.53] and [8.48].

The advantage of modeling by homogenization is that it provides the definitions of the macroscopic parameters, like the rigidity tensor $\boldsymbol{C}$ (see [8.50]), the hydro-mechanical coupling tensor $\boldsymbol{\alpha}$ ([8.51]), and the scalar β (see [8.55]), and the effective diffusion tensor $\boldsymbol{D}^*$ ([8.59]). All these quantities can be calculated, if the microstructure of the porous aquifer is known, by using numerical methods.

8.4. Homogenization of the C-M problem

Let us now assume that the porous medium is subjected to homogeneous solid dissolution. Such a situation corresponds to the long-term (slow) degradation mechanism. In this section, we focus on the analysis of the chemo-mechanical coupling. We are interested in the progressive alteration of the mechanical parameters as a consequence of material dissolution.

8.4.1. *Formulation of the problem at the microscopic scale*

The following two microscopic problems for the non-dimensional variables are formulated in the period domain (Figure 8.3):

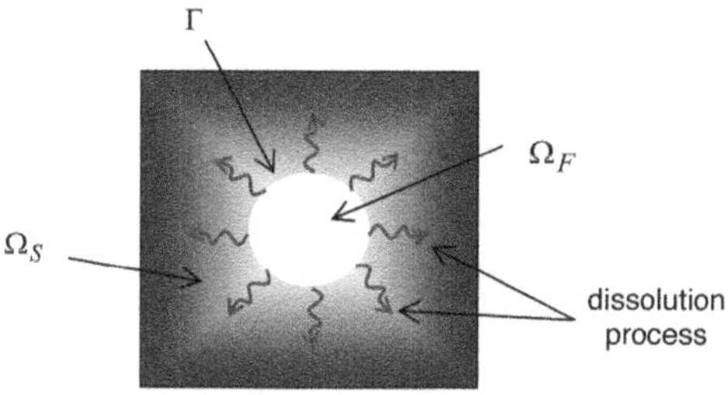

Figure 8.3. *The period with the indicated dissolution process*

– The mechanical "M-problem".

We recall the previously formulated mechanical problem, with $a_{ijkl}(\boldsymbol{y})$ being now the microscopic rigidity tensor depending on the local space variable $\boldsymbol{y}$:

$$\sigma_{Sij,j} = 0 \ in \ \Omega_S \qquad [8.60]$$

$$\sigma_{Sij} = a_{ijkl} e_{kl}(\boldsymbol{u}) \ in \ \Omega_S \qquad [8.61]$$

$$e_{kl} = 1/2(u_{k,l} + u_{l,k})$$

$$\sigma_{Sij} N_j = 0 \ on \ \Gamma \qquad [8.62]$$

– The chemical "C-problem".

The dissolution problem is governed by a single diffusion equation for the concentration in solid (solid fraction) c_S, with one characteristic parameter D_S describing the rate of the process. Such assumption is an evident simplification of the complex chemical reactions taking place at the microscopic scale, and will be used as a first approach to the problem. We further introduce a local dissolution mechanism. This process starts at the interface Γ, and progresses toward the inside of the solid (Figure 8.3):

$$c_S(y,t) = 0 \ in \ \Omega_F \qquad [8.63]$$

$$\frac{\partial c_S}{\partial t} = div(\boldsymbol{D}_S \ grad \, c_S) \ in \ \Omega_S \qquad [8.64]$$

where the boundary and initial conditions are:

$$c_S = 0 \ on \ \Gamma \qquad [8.65]$$

$$N(D_S \ grad \ c_S) = 0 \ on \ \partial\Omega_S \cap \partial\Omega \qquad [8.66]$$

$$c_S(t=0) = 1 \ in \ \Omega_S$$

where $\boldsymbol{D}_S$ is the diffusion tensor related to the evolution of the solid dissolution. Note that we further assume that it is constant but the extension to the case $D_S(c_S)$ is possible.

The dissolution problem is time-dependent, which also makes the associated elasticity problem time-dependent, through the dependence of the local rigidity tensor a_{ijkl} on the concentration in solid (c_S). It is assumed that the constitutive relationship $a_{ijkl}(c_S)$ is known.

8.4.2. *Homogenization*

The homogenization process follows the classical scheme. It can be shown that the results obtained have a similar form to those presented in section 8.3.3. In this case, the macroscopic model is written as:

$$< \frac{\partial \sigma_{Sij}^{(0)}}{\partial x_j} > = 0 \qquad [8.67]$$

where the macroscopic stress tensor is:

$$\sigma_{Sij}^{(0)} = C_{ijkl} \, e_{xkl}(u^{(0)}) \qquad [8.68]$$

The macroscopic tensor C_{ijkl} is the rigidity tensor that can be obtained from the iterative solution of the local boundary value problem. The local boundary value problem is the same as in the case presented in section 8.3.3.2. It should be noted that in the case considered in this section we have to solve the coupled chemo-mechanical problem for each time increment. The two local boundary value problems to be solved simultaneously are:

– The mechanical "M-problem":

$$\frac{\partial}{\partial y_j}(a_{ijkl}(c_S) + a_{ijmn}(c_S) e_{ymn}(\xi^{kl})) = 0 \ in \ \Omega_S \qquad [8.69a]$$

$$(a_{ijkl}(c_S)+(a_{ijmn}(c_S)e_{ymn}(\xi^{kl}))N_j=0 \; on \; \Gamma \qquad [8.69b]$$

$$\xi^{kl} \text{ is periodic} \qquad [8.69c]$$

– The chemical “C-problem”:

$$\frac{\partial c_S}{\partial t}=div(D\, grad\, c_S) \; in \; \Omega_S \qquad [8.70a]$$

$$c_S=0 \; on \; \Gamma \qquad [8.70b]$$

$$N(D\, grad\, c_S)=0 \; on \; \partial\Omega_S \cap \partial\Omega \qquad [8.70c]$$

$$c_S(t=0)=1 \; in \; \Omega_S \qquad [8.70d]$$

The unknown of the problem is the displacement vector $\xi^{kl}(\xi_1^{kl},\xi_2^{kl},\xi_3^{kl})$, which depends on the local space variable y and time t, $\xi^{kl}(y,t)$. Remark that ξ^{kl} is a particular periodic solution corresponding to the unit macroscopic strain tensor $\mathbf{E}^{kl}=(e_k\otimes e_l + e_l\otimes e_k)/2$, where $k, l = 1, 2, 3$. The symbol $\otimes$ means the tensorial product between the unit vectors of the basis. This solution should also verify the zero volume average condition.

Once ξ^{kl} is known, the macroscopic rigidity tensor is defined as the volume average for each time increment as follows:

$$C_{ijkl}=\langle a_{ijkl}(c_S)+a_{ijmn}(c_S)e_{ymn}(\xi^{kl})\rangle \qquad [8.71]$$

Thus, the macroscopic tensor is time-dependent, $\boldsymbol{C}(t)$. In section 8.5 we present an example of the numerical computations of this tensor for a particular microstructure of the porous medium.

8.4.3. *Summary of the macroscopic “C-M model”*

The chemo-mechanical model related to the modification of the microstructure of the porous aquifer is proposed. The model consists of the macroscopic equation of the mechanical equilibrium [8.67] and [8.68] that is coupled with the local model describing the chemical degradation of the mechanical parameters [8.69]–[8.71]. The computations of the evolution of the mechanical parameters can be carried out, if the initial microstructure is given.

It has to be pointed out that the presented model does not take into account the modification of the hydro-mechanical coupling parameters (as proposed in [LEW 12]).

8.5. Numerical computations of the time degradation of the macroscopic rigidity tensor

8.5.1. *Definition of the problem*

Let us analyze the porous medium presenting three-dimensional periodic microstructure, Figure 8.4. The period is a cube of the unit dimension. The three cylinders representing pores have the diameter 0.1. Both domains (fluid Ω_F and solid Ω_S) are connected. The aim is to calculate the macroscopic fourth-order rigidity tensor C that can be converted into the second-order tensor D (6×6) by using the classical conversion rules. The two coupled chemo-mechanical problems [8.69] and [8.70] were solved using the commercial code Comsol Multiphysics.

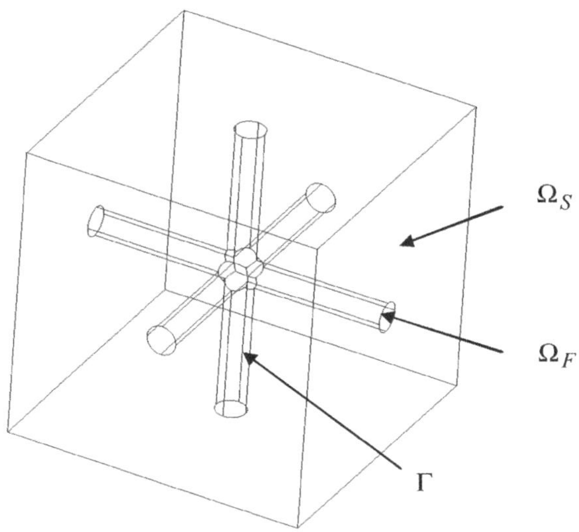

Figure 8.4. *The microstructure considered in the numerical example*

Initially, the (intact) material is locally isotropic with two elasticity parameters: the Young's modulus $E = 1$ and the Poisson's ratio $\nu = 0.2$. As a first approximation, it is assumed that the local constitutive relationship $a_{ijkl}(c_S)$ is such that the only parameter affected by the local variation of the volumetric fraction of the solid is the

Young's modulus. We used the formula of Hashin and Shtrikman (1963) for the upper bound of the effective Young's modulus of a porous material [ROB 02]:

$$\frac{E}{E_S}=\frac{c_S}{1+S(1-c_S)} \qquad [8.72]$$

where S is given as a function of the Poisson's ratio ν_S by the following formula:

$$S=\frac{(1+\nu_S)\times(13-15\nu_S)}{2\times(7-5\nu_S)} \qquad [8.73]$$

and c_S is the solid fraction. In the example, the tensor $\boldsymbol{D}_S$ is assumed isotropic $\boldsymbol{D}_S = \boldsymbol{D}_S \boldsymbol{I}$, where $\boldsymbol{I}$ is the identity matrix and $D_S = 10^{-9}\ m^2/s$.

The period possesses three identical planes of geometry and material symmetry, therefore the rigidity tensor at the macroscopic scale presents cubic symmetry, with three independent elasticity parameters. The following general property holds: $D_{11}=D_{22}=D_{33}$, $D_{12}=D_{13}=D_{23}=D_{21}=D_{31}=D_{32}$ and $D_{44}=D_{55}=D_{66}$. To calculate the components D_{11}, D_{21} and D_{66}, two elementary mechanical problems have to be solved, namely the traction test and the shearing test. We considered the tests:

– traction test with $\mathbf{E}^{11}= (e_1 \otimes e_1 + e_1 \otimes e_1)/2$.

– shearing test with $\mathbf{E}^{12}= (e_1 \otimes e_2 + e_1 \otimes e_2)/2$.

For each test we take into account the transient chemical degradation of the local parameters of the material.

Because of the symmetry, instead of solving the coupled problem ([8.69] and [8.70]) in the full period domain with the periodicity conditions, we solved the problem in the 1/8 of the period (the volume is $0.5 \times 0.5 \times 0.5$) with the appropriate boundary conditions as in [BOR 11]. For example, for the traction tests the boundary conditions on all external boundaries are:

$$u_N = 0 \text{ and } \sigma_T = 0.$$

For the shearing test in the plane (i, j) we have the following conditions:

$u_T = 0$ and $\sigma_N = 0$ on the external boundaries orthogonal to the axis i and j;

$u_N = 0$ and $\sigma_T = 0$ on the external boundaries orthogonal to the axis k;

where u_N and u_T are the displacement in the direction normal to the plane and the displacement in the directions tangent to the plane, respectively. σ_N and σ_T are the normal and tangential stress components, respectively.

8.5.2. *Results and discussion*

In Figure 8.5(a) the result of the computations of the solid concentration (solid fraction) at time t = 1e7 is presented. We can observe the decrease of the solid concentration initiated in the pores. The maximum value in the domain is equal to 1. In Figure 8.5(b) the intensity of the displacement vector $\xi^{11}(\xi_1^{11}, \xi_2^{11}, \xi_3^{11})$ at t = 1e7 for the case of unit macroscopic traction test in the direction 1 is shown. In Figures 8.6(a) and (b) the solid concentration and the intensity of the displacement vector ξ^{11} at t = 1e8 are presented. The maximum value of the concentration in the domain at this time is equal to 0.594. It can be observed that strain is localized in the weak (the most altered) zones.

In Figure 8.7, the three components of the rigidity tensor, C_{11}, C_{12} and C_{66}, as functions of time and as function of average matrix porosity, are plotted. It can be seen that all components of the rigidity matrix are logarithmically decreasing with time. On the other hand, a dramatic decrease of all three components with the increasing average porosity (or decreasing average solid concentration) of the matrix is seen. If we interpret the results in terms of engineering parameters (Figure 8.8), we can see that the Poisson's ratio ν seems to decrease to an asymptotic value equal to 90% of the initial value. Note that it was assumed that the local Poisson's ratio in not altered. The Young's modulus E and the shear modulus G decrease significantly to about 50% of the initial values for the average porosity of the matrix of approximately 30%. The most affected parameter in this case is the shear modulus.

The results of numerical computations using the model obtained by homogenization can be compared with the experiments published in the literature (e.g. [NGU 11]). For example, in Figure 8.9 [NGU 11] the drained bulk modulus and the shear modulus against porosity for the carbonate samples (named by the authors the Euville limestone) are plotted. These parameters were measured in standard triaxial tests. The curves show a decrease of the parameters under the effect of chemical alteration, which manifests itself by the increase of the porosity. It can be seen that these curves agree qualitatively with the computed curves, coming from coupled chemo-mechanical model obtained by homogenization. It has to be pointed out that the microstructure of the carbonate investigated in [NGU 11] is very complex, with bimodal distribution of pore sizes (microporosity and macroporosity). It is believed that the quantitative reproduction of the observed behavior would be possible by following the presented approach, if the real geometry of the microstructure were considered.

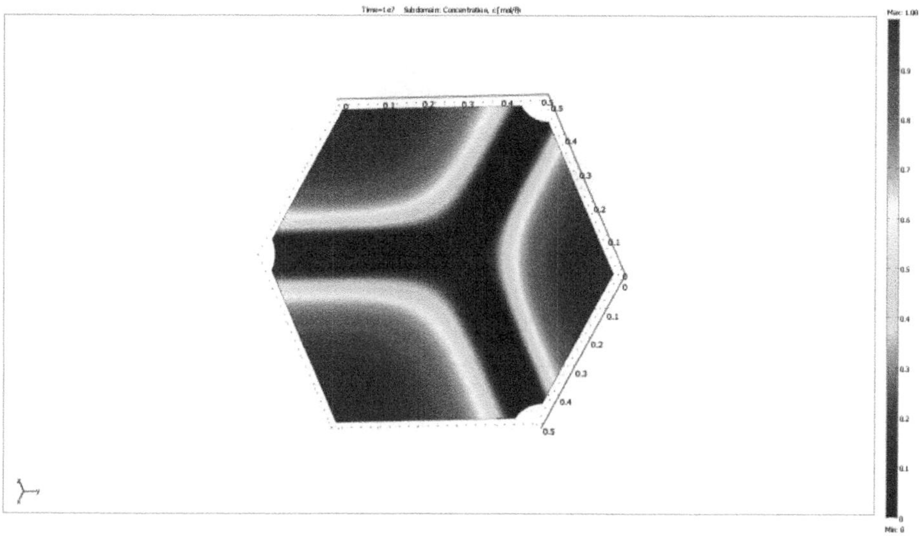

a)

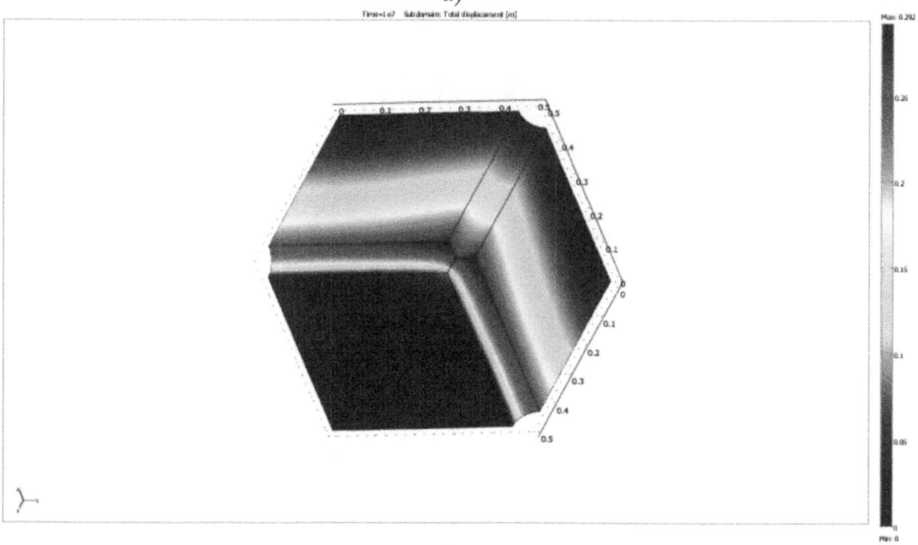

b)

Figure 8.5. *a) The distribution of the solid concentration from 0 to 1 in the 1/8 of the period at time $t=10^7$s. b) The intensity of the displacement vector ξ^{11} ($\xi_1^{11}, \xi_2^{11}, \xi_3^{11}$) in the 1/8 of the period at time $t=10^7$s. The axis 1 is orthogonal to the surface seen in the frontal plane. For a color version of this figure, see www.iste.co.uk/gpc/geomech.zip*

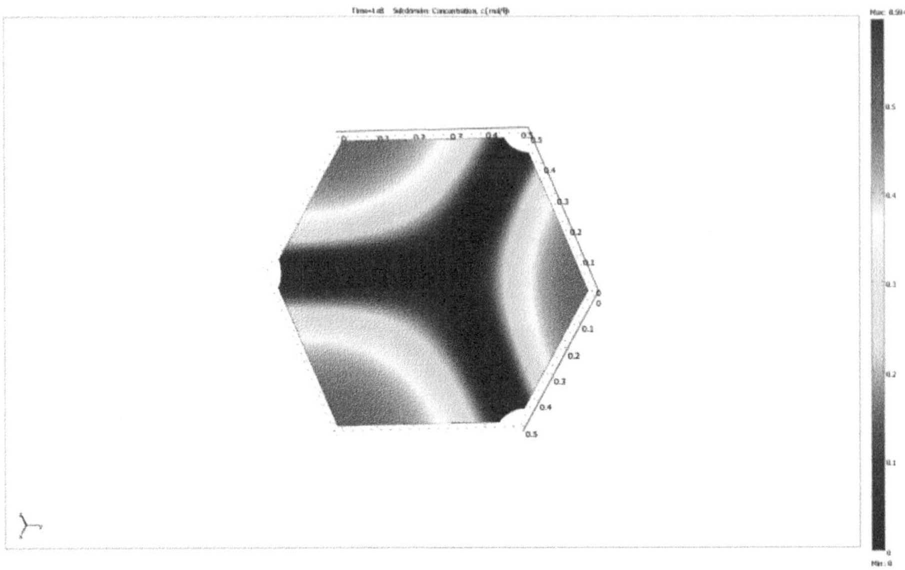

a)

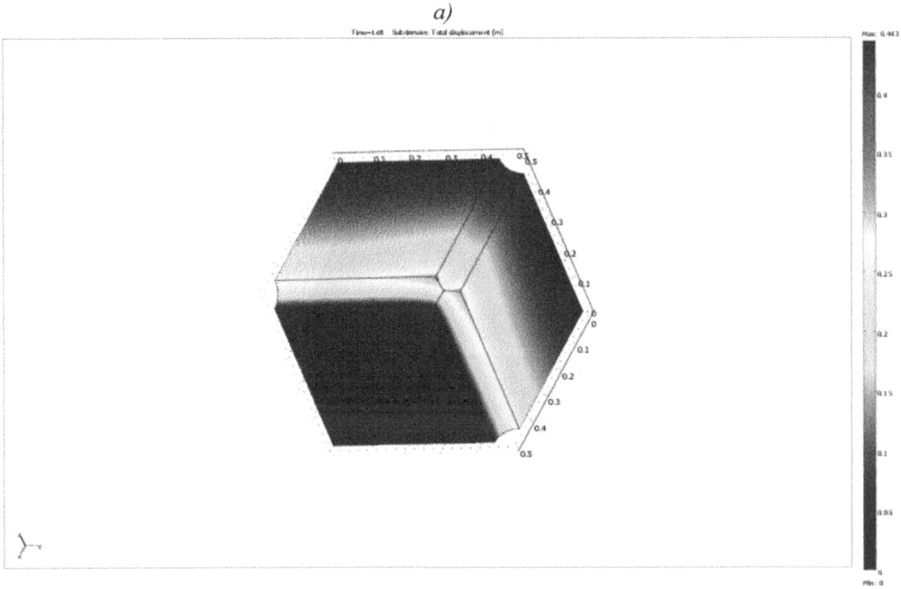

b)

Figure 8.6. *a) The distribution of the solid concentration from 0 to 0.594 in the 1/8 of the period at time t=10^8s. b) The intensity of the displacement vector* ξ^{11} $(\xi_1^{11}, \xi_2^{11}, \xi_3^{11})$ *in the 1/8 of the period at time t=10^8s. The axis 1 is orthogonal to the surface seen in the frontal plane. For a color version of this figure, see www.iste.co.uk/gpc/geomech.zip*

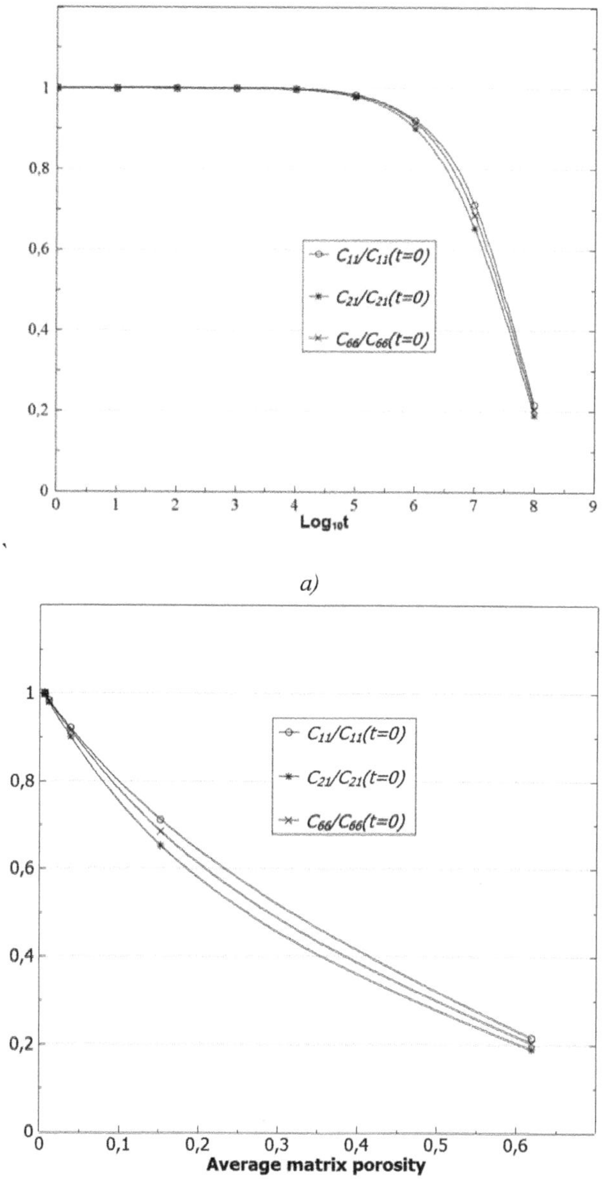

a)

b)

Figure 8.7. *The plots of the three components of the macroscopic rigidity tensor: C_{11}, C_{12}, and C_{66}, as functions of time (a), and as functions of the average matrix porosity (b)*

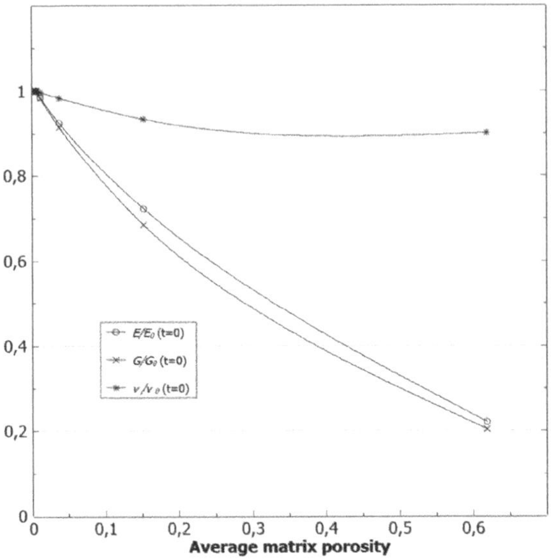

Figure 8.8. *The variation of the three independent engineering parameters of elasticity (E, G, ν) with respect to the initial values (at t = 0) as a function of the average matrix porosity*

8.6. Conclusions

The macroscopic models describing the behavior of a porous saturated aquifer in which the injected CO_2 is migrating are presented. The theoretical framework is the homogenization of periodic structures. This homogenization method is one of the micromechanical approaches that are very well established in the literature. The analysis was focused on the long-term effects, including the chemo-mechanical coupling. The model of chemical degradation and mechanical weakening of the material at long time was proposed. It was shown that the chemo-mechanical coupled computations for a particular three dimensional microstructure can be performed by using the finite elements commercial code Comsol Multiphysics.

The results of the numerical computations of the full macroscopic rigidity tensor showed the degradation with time of all non-zero components of the tensor. Also, the decrease of the rigidity tensor is observed with the decrease of the average solid fraction (or increase of the average porosity). The material of porous aquifer analyzed in the numerical example shows the cubic symmetry at the macroscopic scale (cubic anisotropy). The three (independent) engineering elasticity parameters (E, G, and ν) decrease with time. It can be seen that the proposed degradation mechanism leads to the significant decrease of the Young's modulus and the shear modulus when the average porosity increases from 0 to 30%, while at the same time the Poisson's ratio seems to tend to as asymptotic value of 90% of the initial value.

The comparison between the numerical computations and the experimental data published in the literature showed that the proposed model of chemo-mechanical coupling is able to qualitatively reproduce the curves of the elastic parameters of a rock subjected to the chemical alteration. To enable the quantitative comparison, the computations have to be performed for the real geometry of the microstructure.

Further investigations concerning the chemical modification of the microstructure and its geomechanical effects are currently carried out.

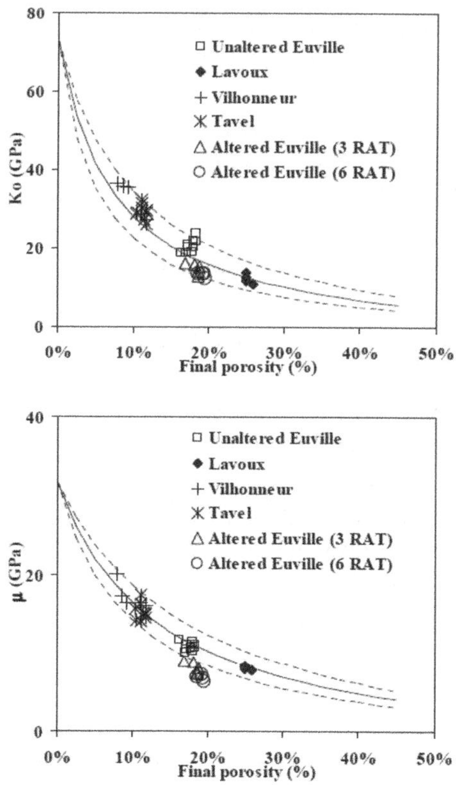

Figure 8.9. *The experimental data of the degradation of the elastic parameters (K, G) of limestone rocks when subjected to chemical alteration. Reproduced after [NGU 11] with permission*

8.7. Acknowledgment

The numerical computations were performed by Artur Lipkowski (Gdansk University of Technology, Poland) in the framework of his Master's thesis prepared at the University of Montpellier 2, Laboratory LMGC (ERASMUS 2012).

8.8. Bibliography

[AUR 05] AURIAULT J.-L., "Upscaling by multiscale asymptotic expansions", in DORMIEUX L., ULM F.-J. (eds), *CISM Lecture 480, Applied Micromechanics of Porous Materials*, Udine 19–23 July 2004, Springer, pp. 3–56, 2005.

[AUR 77] AURIAULT J.-L., SANCHEZ-PALENCIA E., "Etude du comportement macroscopique d'un milieu poreux saturé déformable", *Journal de Mécanique*, vol. 16, no. 4, pp. 576–603, 1977.

[AUR 91] AURIAULT J.-L., "Heterogeneous medium. Is an equivalent description possible?" *International Journal of Engineering Science*, vol. 29, no. 7, pp. 785–795, 1991.

[AUR 96] AURIAULT J.-L., LEWANDOWSKA J., "Diffusion/adsorption/advection macrotransport in soils", *European Journal of Mechanics A/Solids,* vol. 15, no. 4, pp. 681–704, 1996.

[BEN 78] BENSSOUSSAN A., LIONS J.L., PAPANICOLAOU G., *Asymptotic Analysis for Periodic Structures*, North-Holland, Amsterdam, The Netherlands, 1978.

[BER 09] BERNER E., LOMBARD J., "From injectivity to integrity studies of CO_2 geological storage chemical alteration effects on carbonates", *Oil & Gas Science and Technology*, 2009.

[BOR 11] BORNERT M., BRETHEAU T., GILORMINI P. (eds), *Homogénéisation en mécanique des matériaux 1,* Hermes Science Ltd, Paris, 2011.

[EGE 05] EGERMANN P., BEKRI S., VIZIKA O., "An Integrated Approach to Assess the Petrophysical Propertes of Rocks Altered by Rock/Fluid Interactions (CO2 injection)", *Paper SCA Presented at the Society of Core Analysts Symposium*, Toronto, Canada, pp. 21–25, August 2005.

[ENE 84] ENE F., Contribution à l'étude des matériaux composites et leur endommagement, Thèse de Doctorat d'Etat Sciences Mathématiques (Mécanique), UPMC, Paris, 1984.

[GOU 10] GOUZE P., LUQUOT L., "X-ray microtomography characterization of porosity, permeability and reactive surface changes during dissolution", *Journal of Contaminat Hydrology,* 2010.

[IZG 08] IZGEC O., DEMITRAL B., BERTIN H., AKIN S., "CO2 injection into saline carbonate aquifer formations I: laboratory investigation", *Transport in Porous Media*, vol. 72, no. 1, pp. 1–24, 2008.

[LEW 12] LEWANDOWSKA J., AURIAULT J.-L., "Extension of Biot theory to the problem of saturated microporous elastic media with isolated cracks or/and vugs" *International Journal for Numerical and Analytical Methods in Geomechanics*, 2012. Published online in Wiley Online Library (wileyonlinelibrary.com).

[LUQ 09] LUQUOT L., GOUZE P., "Experimental determination of porosity and permeability changes induced by injection of CO_2 into carbonate rocks", *Chemical Geology*, vol. 265, no. 3, pp. 148–159, 2009.

[NGU 11] NGUYEN M.T., BERNER M.T., DORMIUEX L., "Micromechanical modeling of carbonate geomechanical properties evolution during acid gas injection", *The 45th US Rock Mechanics/Geomechanics Symposium*, ARMA, American Rock Mechanics Association, San Francisco, CA, p. 10, June 26–29, 2011.

[NGU 12] NGUYEN M.T., Caractérisation géomécanique de la dégradation des roches sous l'effet de l'injection de gaz acides, PhD Thesis, University of Paris-Est, 2012.

[ROB 02] ROBERTS P., GARBOCZI E.J., "Computation of the linear elastic properties of random porous material with wide variety of microstructure", *Proceedings of the Royal Society London A*, vol. 458, pp. 1033–1054, 2002, doi:10.1098/rspa.2001.0900.

[SAN 80] SANCHEZ-PALENCIA E., *Non-Homogeneous Media and Vibration Theory*, Springer–Verlag, Berlin Heidelberg, New York, 1980.

Chapter 9

Chemoplastic Modeling of Petroleum Cement Paste under Coupled Conditions

9.1. Introduction

In this chapter, we present an elastoplastic model coupled with chemical degradation for a petroleum cement paste. This work is performed in the framework of a feasibility study for the acid gas sequestration in abounded reservoirs. The main objective is to characterize the effects of chemical leaching by acid fluids on the hydromechanical behavior of cement paste and to propose an appropriate constitutive model for the description of this chemical and hydromechanical coupling. For this purpose, a specific cement paste is fabricated under temperature and pressure close to reservoir conditions. The cement paste is subjected to an accelerated chemical leaching process using ammonium nitrate (NH_4NO_3). Triaxial compression tests are performed respectively on sound and degraded samples. It is found that the chemical leaching modifies the microstructure of cement paste with a significant increase of porosity and affects both elastic and plastic properties. The specific fabrication conditions and mineral compositions entrust different microstructure and hydromechanical properties to the petroleum cement: high porosity, strong pressure dependency, and high volumetric compressibility. Further, coupled tests during which the cement paste is subjected to chemical solution injection and deviatoric stress are performed. In this chapter, we propose an elastoplastic model with two yield surfaces, taking into account the effects of

Chapter written by Jian Fu SHAO, Y. JIA, Nicholas BURLION, Jeremy SAINT-MARC and Adeline GARNIER.

chemical leaching. The proposed model is applied to the simulation of the coupled chemo-mechanical tests.

9.2. General framework for chemo-mechanical modeling

Here, the cement paste is considered as a porous material composed of the connected pore space and the solid skeleton. According to experimental investigations [YUR 11a, YUR 11b], the basic mechanical behavior of studied cement paste can be characterized by elastic–plastic deformation with mechanical damage. However, the mechanical damage is significant only under very low confining stresses and its effect on elastic properties is quite small. Therefore, putting the emphasis on the chemical degradation, the basic mechanical behavior of cement paste will be described by an elastoplastic model by neglecting induced mechanical damage. Furthermore, it appears reasonable to assume an isotropic behavior for the cement paste.

Under the chemically sound condition, the solid constituents of cement paste and the substances constituting the pore fluid (cations, anions, and water) are in chemical equilibrium. ϕ_r denotes the connected porosity of the sound material. Due to the change of environmental conditions, the chemical equilibrium is broken leading to the dissolution of solid constituents and the diffusion of solved species inside the interstitial fluid. The main chemical reactions related to the dissolution of cement paste have been widely discussed by various authors [BER 88, ADE 92, GER 96]. Since the kinetics of leaching in natural solutions such as pure water may be slow, to perform laboratory investigations during a reasonable duration, different kinds of acceleration methods have been proposed. Among these, the ammonium nitrate solution has been largely used in a series of works [CAR 96, CAR 97, HEU 01]. The equivalence of the leaching process in cement-based materials exposed to deionized water and to ammonium nitrate solutions is shown in [CAR 97] by means of chemical analyses of standard and accelerated leached cement paste. According to these investigations, when the cement-based materials are exposed to ammonium nitrate solution, the dissolution of portlandite, ettringite, and C-S-H phases occurs at different time sequences. However, it is found that the dissociation of portlandite represents the most important effect on the mechanical behavior of cement paste. The dissociation of calcium ions from the solid skeleton during the chemical leaching process leads to the increase of pore space and thus the increase of the total connected porosity. From the mechanical point of view, the increase of porosity due to chemical leaching can be considered as a damage process, which is described by a scalar valued chemical damage variable [GER 98, KUH 04, BEL 03]:

$$d_c = \phi - \phi_r \quad [9.1]$$

We present here the general framework for the formulation of elastoplastic models by taking into account chemical leaching. The state variables are the elastic strain tensor ε_{ij}^{e}, the plastic strain tensor ε_{ij}^{p} and the chemical damage variable d_c. The assumption of small strains is adopted throughout this chapter and the strain increment partition rule is applied:

$$d\varepsilon_{ij} = d\varepsilon_{ij}^{e} + d\varepsilon_{ij}^{p} \quad [9.2]$$

As mentioned above, assuming an isotropic behavior of cement paste, the thermodynamic potential can be expressed as follows:

$$\Psi(\varepsilon^{e}, \gamma_p, d_c) = \frac{1}{2}\varepsilon^{e} : \mathbb{C}(d_c) : \varepsilon^{e} + \Psi_p(\gamma_p, d_c) \quad [9.3]$$

The fourth-order tensor $\mathbb{C}$ is the elastic stiffness of chemically damaged material. The function Ψ_p is the locked energy for plastic hardening. The standard derivation of the thermodynamic potential yields the state equation:

$$\boldsymbol{\sigma} = \frac{\partial \Psi}{\partial \boldsymbol{\varepsilon}^{e}} = \mathbb{C}(d_c) : (\boldsymbol{\varepsilon} - \boldsymbol{\varepsilon}^{p}) \quad [9.4]$$

The effective elastic stiffness tensor of isotropic damaged material reads:

$$\mathbb{C}(d_c) = 2\mu(d_c)\mathbb{K} + 3k(d_c)\mathbb{J} \quad [9.5]$$

where $k(d_c)$ and $\mu(d_c)$ are the bulk and the shear moduli of damaged material, respectively. The two fourth-order isotropic tensors are defined by:

$$\mathbb{J} = \frac{1}{3}\boldsymbol{\delta} \otimes \boldsymbol{\delta}, \quad \mathbb{K} = \mathbb{I} - \mathbb{J} \quad [9.6]$$

where $\boldsymbol{\delta}$ is the second-order unit tensor. The thermodynamic force associated with the plastic hardening variable is given by:

$$A_p = \frac{\partial \Psi}{\partial \gamma_p} = \frac{\partial \Psi_p(\gamma_p, d_c)}{\partial \gamma_p} \quad [9.7]$$

In the case of isotropic hardening, this force variable defines the current size of elastic convex of damaged material. The rate form of the constitutive equation [9.15] can easily be written as:

$$\dot{\boldsymbol{\sigma}} = \mathbb{C}(d_c):(\dot{\boldsymbol{\varepsilon}} - \dot{\boldsymbol{\varepsilon}}^p) + \mathbb{C}'_c\,(d_c):(\boldsymbol{\varepsilon} - \boldsymbol{\varepsilon}^p)\dot{d}_c \qquad [9.8]$$

The fourth-order tensor $\mathbb{C}'_c\,(d_c)$ is the derivative of the elastic stiffness tensor with respect to the chemical damage variable. We can see that the formulation of the constitutive model needs to define a specific plastic flow rule and a chemical damage evolution law to determine the evolutions of plastic strain and chemical damage.

9.2.1. *Phenomenological chemistry model*

In the present work, the chemical damage is defined as the variation of porosity due to the dissolution of solid calcium in the cement paste. By assuming an instantaneous local chemical equilibrium, the kinetics of dissolution is described by the local calcium mass conservation as follows [GER 98, KUH 04, BEL 03]:

$$\frac{\partial\left(\phi Ca^{++}\right)}{\partial t} = -\operatorname{div}(\vec{\Phi}_{Ca^{++}}) + M^0_{C_a^{solid} \to Ca^{++}} \qquad [9.9]$$

where ϕ is the total connected porosity, $\vec{\Phi}_{Ca^{++}}$ is the flux vector of calcium ions in pore fluid, Ca^{++} is the calcium concentration in interstitial fluid, and $M^0_{Ca^{solid} \to Ca^{++}}$ is the rate of calcium (Kg/s/m^3) dissolved from the solid skeleton into the interstitial fluid. Further, the calcium concentration in the skeleton Ca^{solid} can be related to that in the interstitial fluid as follows:

$$M^0_{Ca^{solid} \to Ca^{++}} = -\frac{\partial Ca^{solid}}{\partial t} = -\frac{\partial f\left(Ca^{++}\right)}{\partial t} \qquad [9.10]$$

The function $f(Ca^{++})$ is determined from the experimental chemical equilibrium curve. Note that the porosity variation can be calculated directly from the quantity of calcium dissolved from the solid skeleton, which is a function of the calcium concentration in pore fluid. The flux vector of calcium ions is governed by Fick's law:

$$\vec{\Phi}_{Ca^{++}} = -D(Ca^{++})\vec{\nabla}Ca^{++} \qquad [9.11]$$

where $D(Ca^{++})$ denotes the diffusion coefficient dependent on the calcium concentration in pore fluid. Then, by combining the calcium mass balance equation and Fick's law, we obtain the following calcium diffusion equation:

$$\left[\frac{\partial \phi}{\partial Ca^{++}} Ca^{++} + \phi\left(Ca^{++}\right) + \frac{\partial Ca^{solid}}{\partial Ca^{++}}\right] \cdot \frac{\partial Ca^{++}}{\partial t} = \operatorname{div}\left[D(Ca^{++}) \cdot \vec{\nabla} Ca^{++}\right] \quad [9.12]$$

It is important to point out that in the first part of equation [9.12], the term of calcium dissolution is generally dominant with respect to the other terms. Therefore, the first two terms inside the bracket can be neglected. After this simplification, the kinetics of calcium leaching is governed by the following simplified diffusion equation:

$$\frac{\partial Ca^{solid}}{\partial Ca^{++}} \cdot \frac{\partial Ca^{++}}{\partial t} = \operatorname{div}\left[D(Ca^{++}) \cdot \vec{\nabla} Ca^{++}\right] \quad [9.13]$$

We can see that the calcium leaching is considered here as a diffusion process which is controlled by the chemical equilibrium between the calcium concentration in solid skeleton Ca^{solid} and that in interstitial fluid Ca^{++}. Based on experimental data and previous works [BER 88, GER 96], the following empirical chemistry model is adopted:

$$Ca^{solid} = f\left(Ca^{++}\right) = a - b(Ca^{++})^2 + cCa^{++} - \left[\frac{e}{1+\left(\frac{Ca^{++}}{x_2}\right)^n} + \frac{h}{1+\left(\frac{Ca^{++}}{x_1}\right)^m}\right] \quad [9.14]$$

where a, b, c, e and h are the parameters of the chemistry model which depend on the proportion of calcium in the cement paste, and m and n are the controlling parameters that can vary between 1 and 100. The parameters x_1, x_2 represent the characteristic values of the calcium concentration in pore fluid, separating the different stages of chemical leaching process. From the experimental results by [BER 88], the parameters used in the chemistry model are related to the portlandite content (S_{por}) and the total calcium content (S_{tot}) by [GÉR 96]:

$$e = S_{por}, \quad h = 0{,}565(S_{tot} - S_{por}), \quad a = S_{por} + h$$
$$b = \frac{(S_{tot} - S_{por} - h)}{400}, \quad c = \left(\frac{(S_{tot} - S_{por} - h)}{20}\right) + 20b \quad [9.15]$$

The values of the chemical parameters are given in Table 9.3. It is now possible to determine the derivative of the chemical equilibrium curve:

$$g(Ca^{++}) = \frac{\partial Ca^{solid}}{\partial Ca^{++}} = -2bCa^{++} + c$$
$$+\left[\frac{1}{x_2^n}\frac{e\cdot n\cdot (Ca^{++})^{(n-1)}}{\left(1+\left(\frac{Ca^{++}}{x_2}\right)^n\right)^2} + \frac{1}{x_1^m}\frac{h\cdot m\cdot (Ca^{++})^{(m-1)}}{\left(1+\left(\frac{Ca^{++}}{x_1}\right)^m\right)^2}\right] \quad [9.16]$$

The calcium leaching modifies pore networks and calcium concentration in pore fluid. As a consequence, the diffusion coefficient is affected by the chemical leaching. In a simplified form, the diffusion coefficient can be expressed as a function of the internal variable Ca^{++} only. Based on experimental data and the previous works in [GÉR 96], the following relation is used:

$$D = D_s\left(\frac{D_0}{D_s}\right)^{\frac{\beta V_{por}^d + \alpha V_a^d}{V_{por}^i + V_a^i}}, \quad \alpha = (1-\beta)\frac{V_{por}^i}{V_a^i} + 1, \quad 0 < \beta < \frac{V_a^i + V_{por}^i}{V_{por}^i} \quad [9.17]$$

The parameters D_0 and D_s are, respectively, the initial value for sound material and the asymptotic value for fully leached diffusion coefficient. V_{por} and V_a are, respectively, the volume fraction of portlandite and that of other hydrates. The indexes d and i, respectively, designate the leached and initial samples. The coefficient β is introduced to account for the effect of portlandite dissolution. Moreover, the diffusion coefficient may also be sensitive to other factors, such as fluid pressure, temperate, and chemical nature of interstitial solution. As far as analytical solutions are not available to determine such influences, a simplified empirical method consists in introducing a multiplier coefficient λ_t so that the actual diffusion is defined by $\lambda_t D$. The value of this multiplier coefficient can be determined from experimental data obtained in different conditions.

Further, based on the assumption of homogeneous molar volume within the skeleton and previous works in [KUH 04], the chemical damage related to the change of porosity can be determined as follows:

$$\phi - \phi_r = \frac{M}{\rho}\left[S_{tot} - f\left(Ca^{++}\right)\right] \quad [9.18]$$

The parameters M and ρ are respectively, the average molar mass and density while S_{tot} denotes the total content of calcium in the skeleton. The relationship between the porosity and the calcium concentration in pore fluid is shown in Figure 9.1.

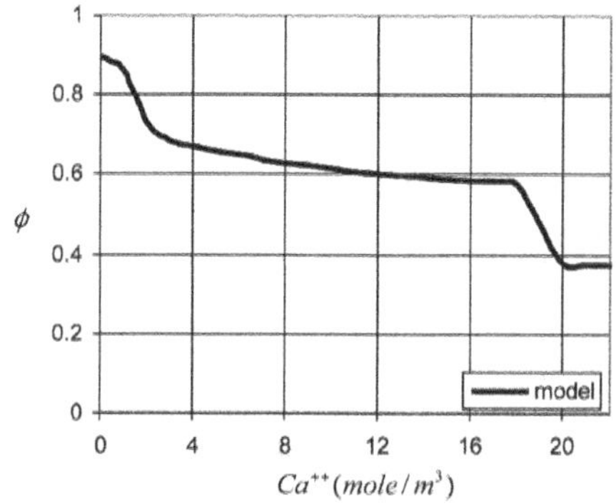

Figure 9.1. *Relationship between total porosity and calcium concentration in pore fluid*

9.3. Specific plastic model for petroleum cement paste

As mentioned above, the basic mechanical behavior of the petroleum cement paste is characterized by the strong pressure dependency and high plastic compressibility. An elastoplastic model with two yield mechanisms is proposed to respectively describe plastic shearing and pore collapse. Under low confining pressures, the plastic shearing is dominant and controls the failure process of material. However, under higher confining pressures, the plastic deformation is significantly controlled by the pore collapse leading to large volumetric compaction. The total plastic strain is divided into a pore collapse component $d\varepsilon_{ij}^{c}$ and a shear component $d\varepsilon_{ij}^{s}$:

$$d\varepsilon_{ij}^{p} = d\varepsilon_{ij}^{c} + d\varepsilon_{ij}^{s} \qquad [9.19]$$

Both the elastic and plastic properties are affected by the chemical damage. However, the mechanical damage of material is neglected.

9.3.1. *Elastic behavior*

The elastic properties of cement paste are affected by the chemical leaching due to the increase of porosity. Based on the micromechanical analysis of porous materials using Mori-Tanaka [MOR 73], the elastic moduli are functions of chemical damage as follows:

$$k(d_m, d_c) = \frac{k_0}{1+\alpha_1 d_c}, \quad \mu(d_m, d_c) = \frac{\mu_0}{1+\alpha_2 d_c} \qquad [9.20]$$

where k_0 and μ_0 are the initial bulk and shear moduli of sound material, respectively, α_1 and α_2 are two parameters defining the progressive deterioration of elastic properties due to chemical leaching.

9.3.2. *Plastic pore collapse model*

The microstructure of cement paste is complex. However, as a first approximation in view of macroscopic modeling, an idealized morphology is adopted. The cement paste is roughly represented by a porous medium constituted by a solid matrix containing connected spherical pores. The closed porosity is included in the solid matrix. With this simplification, the micromechanics-based plastic model proposed by Gurson [GUR 77] is considered as the reference model for this class of porous materials. This model is deduced by a limit analysis method of a porous material with von-Mises type perfect plastic solid matrix and spherical pores. Gurson's model has been extended and improved by various authors to account for the pore geometry and the plastic compressibility of the solid matrix. This model has also been used for plastic modeling of porous chalks [XIE 06].

For the sake of simplicity and considering the sound physical background of Gurson's model, we here adopt the yield function of this model for the plastic pore collapse of cement paste:

$$f_c = \frac{q^2}{\bar{\sigma}^2} + 2\phi \cosh\left(\frac{3p}{2\bar{\sigma}}\right) - 1 - \phi^2 = 0 \qquad [9.21]$$

The stress variables, p and q, denote the mean and deviatoric stresses, respectively:

$$p = \frac{1}{3}\sigma_{kk}, \quad q = \sqrt{\frac{3}{2} s_{ij} s_{ij}}; \quad s_{ij} = \sigma_{ij} - \frac{1}{3}\sigma_{kk}\delta_{ij} \qquad [9.22]$$

In [9.21], $\bar{\sigma}$ represents the current yield stress of the solid matrix. It is assumed that the yield stress of the solid matrix is a function of volumetric plastic strain. Based on experimental data obtained in hydrostatic compression tests, the following hardening law is proposed:

$$\bar{\sigma} = \bar{\sigma}_0 \left[1 + a(\varepsilon_{kk}^c)^n e^{b\varepsilon_{kk}^c} \right] \quad [9.23]$$

where $\bar{\sigma}_0$ defines the initial yield stress of solid matrix. Three parameters b, n and a control the kinetics of plastic hardening and can be identified from hydrostatic compression test. With the assumption that the plastic compression of a solid matrix may be neglected, the porosity variation due to plastic deformation is given by:

$$d\phi = (1-\phi)\, d\varepsilon_{kk}^c \quad [9.24]$$

Further, an associated plastic flow rule is proposed for the pore collapse mechanism. The plastic potential is written as:

$$Q_c = \frac{q^2}{\bar{\sigma}^2} + 2\phi \cosh\left(\frac{3p}{2\bar{\sigma}}\right) \quad [9.25]$$

In the absence of plastic shearing, the plastic flow rule can be written as:

$$d\varepsilon_{ij}^c = d\lambda_c \frac{\partial Q_c}{\partial \sigma_{ij}} \quad [9.26]$$

where $d\lambda_c$ is the plastic multiplier which can be determined from the classical plastic consistency condition.

On the other hand, comparing the stress strain curves in hydrostatical compression tests between the sound and degraded samples [YUR 11a, YUR11b], it is found that the yield threshold of pore collapse is significantly affected by the chemical damage. Thus, to account for the chemical degradation, the hardening law for the pore collapse mechanism is modified as follows:

$$\bar{\sigma} = \bar{\sigma}_0(d_c)\left[1 + a(\varepsilon_{kk}^c)^n e^{b\varepsilon_{kk}^c} \right], \quad \bar{\sigma}_0(d_c) = \frac{\bar{\sigma}_{s0}(d_c = 0)}{1 + \alpha_4 d_c} \quad [9.27]$$

The parameter α_4 defines the chemical effect on the plastic pore collapse mechanism. Accordingly, without the activation of plastic shearing and under a stress-controlled loading, the plastic strain rate is determined using the plastic consistency condition:

$$d\varepsilon_{ij}^c = d\lambda_c \frac{\partial Q_c}{\partial \sigma_{ij}} = \left(\frac{\dfrac{\partial f_c}{\partial \sigma_{ij}} d\sigma_{ij} + \dfrac{\partial f_c}{\partial d_c} dd_c}{h_c} \right) \frac{\partial Q_c}{\partial \sigma_{ij}} \quad [9.28]$$

where h_c is the plastic hardening modulus for pore collapse mechanism. It is clear that the second term in the bracket represents the contribution of chemical damage to plastic deformation of cement paste, which can occur even in the absence of stress variation.

9.3.3. *Plastic shearing model*

According to the experimental investigation [YUR 11a, YUR11b], the mechanical behavior of petroleum cement paste exhibits a strong pressure dependency. With the increase of confining pressure, the mechanical behavior clearly changes from a brittle regime to that of a ductile. The shearing failure surface cannot be described by a classic linear Mohr–Coulomb-type criterion. In view of these pieces of evidence, a curved yield surface is proposed for the yield function of plastic shearing:

$$f_s = \left(\frac{q}{p_r}\right)^{3/2} + \alpha_s A\left(\frac{p}{p_r} - C_s\right) = 0 \qquad [9.29]$$

The parameter C_s represents the hydrostatic tensile strength of the sound material. p_r is a normalizing coefficient of stresses and taken here as the unit stress. The parameter A controls the opening of elastic convex and is related to the current values of cohesion and frictional angle. For reasons of simplicity, the influence of the Lode angle is neglected here. The evolution of the shearing surface is controlled by the hardening function α_s, depending on the plastic distortion γ_s as follows:

$$\alpha_s = \frac{\gamma_s}{B + \gamma_s} \qquad [9.30]$$

$$d\gamma^s = \sqrt{\frac{2}{3} de_{ij}^s de_{ij}^s}, \quad de_{ij}^s = d\varepsilon_{ij}^s - \frac{1}{3} d\varepsilon_{kk}^s \delta_{ij} \qquad [9.31]$$

The parameter B controls the kinetics of plastic hardening. Further, as for most geomaterials, a non-associated flow rule is needed for the plastic shearing mechanism to better capture the transition between plastic compressibility and dilatancy. Based on experimental data [YUR 11a, YUR11b] and inspired by the plastic model proposed by [PIE 88] for concrete, the following plastic potential is used:

$$Q_s = q + \eta_s \bar{I} \ln\left(\frac{\bar{I}}{I_0}\right) = 0, \quad \bar{I} = p - p_r C_s \qquad [9.32]$$

The variable I_0 defines the intersection point between the potential surface and the axis p. The transition from plastic compressibility to dilatancy occurs on the boundary defined by the condition $\partial Q_s / \partial p = 0$. In this model, it is assumed that the transition boundary can be described by a straight line:

$$\bar{f}_{cd} = \sigma_{eq} - \eta_s \bar{I} = 0 \qquad [9.33]$$

The parameter η_s defines the slope of the transition line in $q - p$ plane. In the absence of pore collapse, the plastic flow rule is written as:

$$d\varepsilon_{ij}^{s} = d\lambda_s \frac{\partial Q_s}{\partial \sigma_{ij}} \qquad [9.34]$$

The plastic multiplier can be determined by the plastic consistency condition.

The comparison of failure surface between sound and chemically damaged samples is shown in Figure 9.2. We can see that the mechanical strength of material significantly decreases due to chemical leaching. As a consequence, the parameter controlling the failure surface, A, is expressed as a function of chemical damage as follows:

$$A(d_c) = \frac{A_0(d_c = 0)}{1 + \alpha_3 d_c} \qquad [9.35]$$

where α_3 is a parameter of the model for chemoplastic coupling. By taking into account the influence of chemical damage on failure surface, in the absence of plastic pore collapse and under a stress-controlled loading, the plastic strain rate is given by:

$$d\varepsilon_{ij}^{s} = d\lambda_s \frac{\partial Q_s}{\partial \sigma_{ij}} = \left(\frac{\frac{\partial f_s}{\partial \sigma_{ij}} d\sigma_{ij} + \frac{\partial f_s}{\partial d_c} dd_c}{h_s} \right) \frac{\partial Q_s}{\partial \sigma_{ij}} \qquad [9.36]$$

where h_s is the plastic hardening modulus for deviatoric shearing mechanism. Again, the second term in the bracket represents the contribution of chemical damage to plastic deformation of cement paste, which can occur even in the absence of stress variation.

During a general loading process, two plastic mechanisms can be activated either separately or simultaneously. Four distinct constitutive regimes can be identified. A specific algorithm is needed for the numerical integration of the chemoplastic model. Details are not given in this chapter.

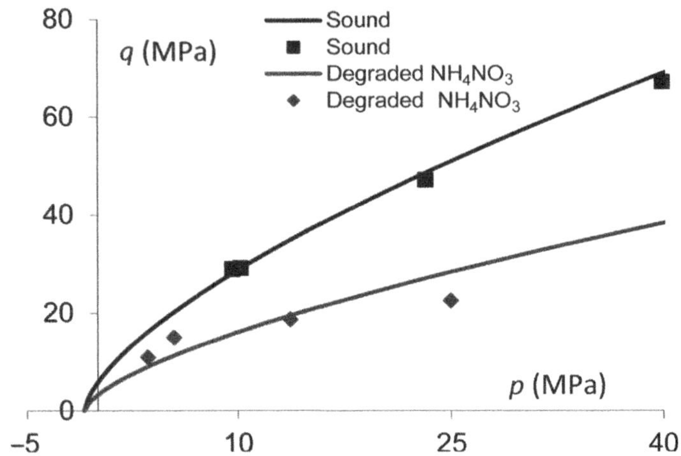

Figure 9.2. *Failure surfaces for intact and damaged cement paste samples*

9.4. Validation of model

The proposed model contains a number of parameters for both elastic and plastic strains. The general strategy for parameter identification is as follows: all the mechanical parameters are first identified from hydrostatic and triaxial compression tests performed on sound cement paste samples. Then by comparing elastic modulus, hydrostatic yield stress, and deviatoric failure surface between sound and degraded materials, the parameters for chemical–mechanical coupling are identified. In Table 9.1, the variation of porosity due to chemical leaching is presented.

Initial porosity of sound sample (%)	Porosity of completely damaged sample (%)	Chemical damage (%)
41.7	67.1	25.4

Table 9.1. *Modification of porosity of cement paste leached by NH_4NO_3*

The parameters obtained are verified by the simulation of mechanical tests on sound and degraded samples. In Figures 9.3 and 9.4, two examples of triaxial compression tests are presented.

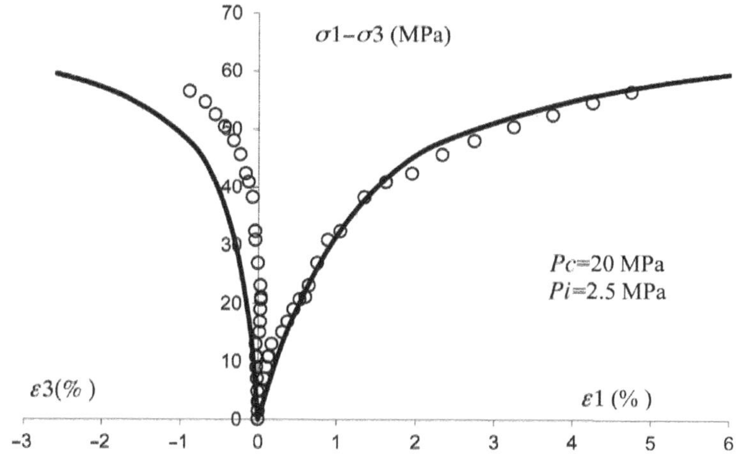

Figure 9.3. *Simulation of a triaxial compression test with 20 MPa confining pressure on sound sample*

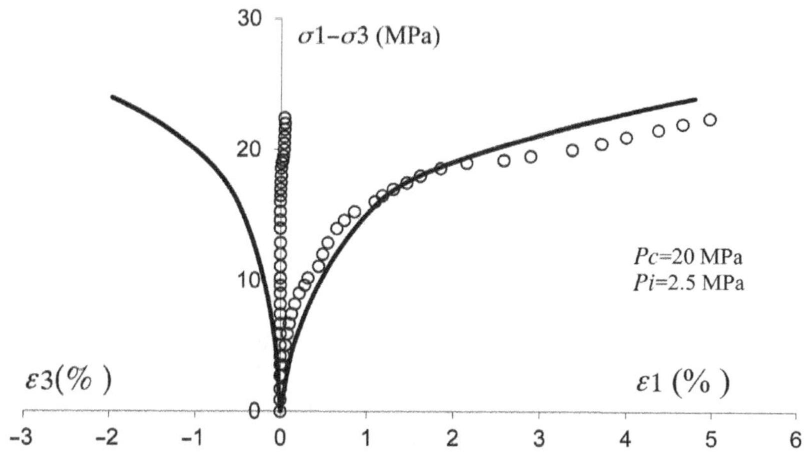

Figure 9.4. *Simulation of a triaxial compression test with 20 MPa confining pressure on degraded sample*

In this chapter, the emphasis is put on the validation of the proposed model for coupled mechanical and chemical tests. In such tests, the cylinder sample is first subjected to a triaxial compression state with selected confining pressure and deviatoric stress. Then, the chemical attack fluid is injected from the bottom surface of the sample. We measure the evolution of axial and radial strains during the injection. The chemical solution used is NH_4NO_3 and the test conditions are summarized in Table 9.2.

Confining pressure (MPa)	Deviatoric stress (MPa)	Injection pressure (MPa)	Drainage pressure (MPa)
$Pc = 3$	15	2.5	0.5
$Pc = 10$	22		

Table 9.2. *Test conditions for coupled tests with NH_4NO_3 solution*

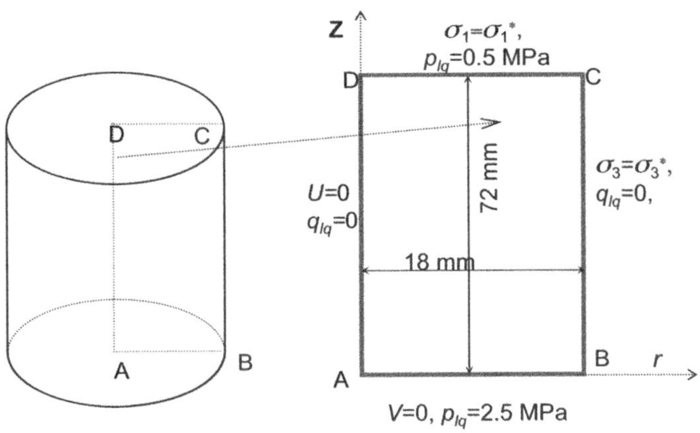

Figure 9.5. *Geometry of cylinder sample and boundary conditions in coupled test*

Note that the leached zone is propagating into the sample during the injection of attack fluid and the tested sample can no longer be seen as a uniform volume element. The sample should be considered as a small structure with prescribed initial and boundary conditions. The geometry of sample and main boundary conditions are shown in Figure 9.5. Due to the symmetry of the problem, only one half part of the sample is considered and meshed. The symmetric axis (AD) and the lateral surface (BC) are impermeable while constant pressures are prescribed respectively on the bottom and top surfaces (AB and CD). Further, the horizontal displacement is blocked on AD and the vertical displacement is blocked on AB. The uniform confining pressure is applied on the lateral surface BC and the deviatoric stress is applied on CD. The typical values of parameters used in the phenomenological chemical model are given in Table 9.3.

S_{tot}	S_{por}	x_1	x_2	n	m
15,278 mol/ m^3	4,427 mol/ m^3	1.5	19	70	5
D_0	D_s	β	V_{por}	V_{CSH}	$V_{Aft+Afm}$
$3 \cdot 10^{-12}\, m^2/s$	$400 \cdot 10^{-12}\, m^2/s$	2.3	0.18	0.40	0.14
M/ρ					
$3.5 \cdot 10^{-5}\, m^3/mol$					

Table 9.3. *Typical values of parameters used in chemical model*

The present chemo-mechanical model is based on the following assumption: the chemically degraded microstructure state of material is the same using different attack fluids while the kinetics to reach such a state depends on the chemical nature of attack fluids. The influence of chemical nature of attack fluid on the degradation kinetics is represented by the multiplier of diffusion coefficient noted by λ_t, such as:

$$\bar{D} = \lambda_t D = \lambda_t D_s \left(\frac{D_0}{D_s} \right)^{\frac{\beta V^d_{por} + \alpha V^d_a}{V^i_{por} + V^i_a}} \qquad [9.37]$$

Further, it is also found that the diffusion coefficient is dependent on applied stresses and in particular on confining pressure. Based on experimental data on the time needed to obtain a complete degradation state of cement paste [CAR 97], the following values are adopted.

Test	Confining pressure	λ_t
NH_4NO_3	3 MPa	50,000
	10 MPa	15,000

Table 9.4. *Multiplier of diffusion coefficient for confining pressures*

In Figures 9.6 and 9.7, the numerical results for two coupled tests with NH_4NO_3 solution are presented. Generally, there is a good agreement between the numerical results and experimental data. The axial strain progressively increases with the propagation of leached zone inside the sample. Under low confining pressure, we can observe an accelerated evolution of axial strain at the final stage of test, leading to failure of sample. However, under high confining pressure, it seems that the axial strain reaches an asymptotic value with significant compaction of sample. Further, it is clear that the kinetics of chemical degradation increases with the decrease of confining pressure. This phenomenon is related to the fact that under higher confining pressure, the chemical diffusion coefficient is reduced by compressive stress.

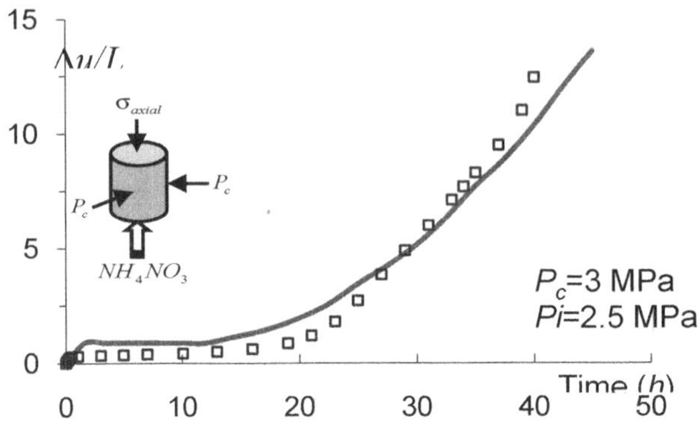

Figure 9.6. *Evolution of axial strain with time for coupled test under confining pressure 3 MPa with injection of NH_4NO_3*

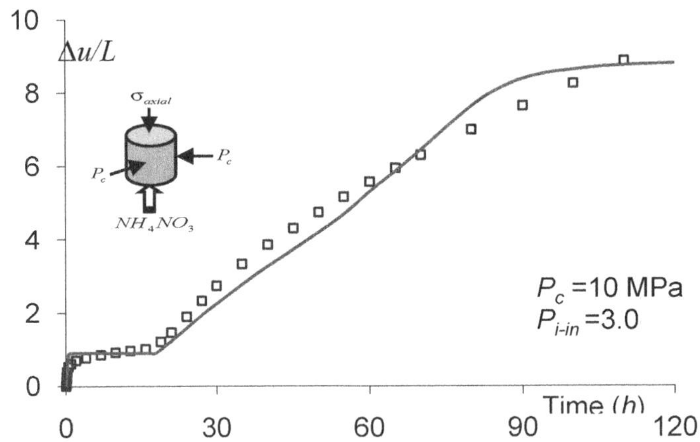

Figure 9.7. *Evolution of axial strain with time for coupled test under confining pressure 10 MPa with injection of NH_4NO_3*

9.5. Conclusions and perspectives

Based on relevant experimental data and with the help of a phenomenological chemical model, a general framework is proposed for constitutive modeling of coupled chemo-mechanical behavior in cement-based materials. The chemical damage is identified as the increase of connected porosity by dissolution of calcium ions from solid cement skeleton. The chemical dissolution is described by a phenomenological chemical model, which gives the relationship between the change of porosity and calcium concentration in interstitial fluid. The chemical damage affects both elastic and plastic properties. For the petroleum cement paste studied here, a simplified version is used by neglecting the mechanical damage. The plastic deformation is described by two complementary mechanisms: plastic shearing and pore collapse. The parameters of the model can be determined from uncoupled triaxial compression tests performed respectively on sound and leached samples. The comparison between numerical simulations and experimental data for uncoupled tests has shown that the proposed model is able to describe the basic mechanical behaviors of the cement paste. The proposed model also correctly describes the main features observed in coupled tests. The influence of chemical nature of the attack fluid on the kinetics of chemical degradation is taken into account through the diffusion coefficient. However, some quantitative scatters have been obtained between numerical results and experimental data in some cases. Further improvements could be envisaged by coupling micromechanical analysis and phenomenological approaches to better describe evolutions of microstructure of cement paste due to chemical degradation and consequences on macroscopic properties.

9.6. Bibliography

[ADE 92] ADENOT F., BUIL M., "Modelling of the corrosion of the cement paste by deionized water", *Cement and Concrete Research,* vol. 22, no. 4, pp. 489–495, 1992.

[BEL 03] LE BELLÉGO C., GÉRARD B., PIJAUDIER-CABOT G., DUBÉ J.-F., MOLEZ L., "Coupled mechanical and chemical damage in calcium leached cementitious structures", *Journal of Engineering Mechanics,* vol. 129, no. 3, pp. 333–341, 2003.

[BER 88] BERNER U., "Modelling the incongruent dissolution of hydrated cement minerals", *Radiochimica Acta,* vol. 44–45, pp. 387–393, 1988.

[BER 92] BERNER U., "Evolution of pore water chemistry during degradation of cement in a radioactive waste repository environment", *Waste Management,* vol. 12, pp. 201–219, 1992.

[CAR 96] CARDE C., FRANÇOIS R., TORRENTI J.M., "Leaching of both calcium hydroxide and C-S-H from cement paste: modeling the mechanical", *Cement and Concrete Research*, vol. 26, no. 8, pp. 1257–1268, 1996.

[CAR 97] CARDE C., ESCADEILLAS G., FRANCOIS R., "Use of ammonium nitrate solution to simulate and accelerate the leaching of cement pastes due to deionized water", *Magazine of Concrete Research,* vol. 49, pp. 295–301, 1997.

[GÉR 96] GÉRARD B., Contribution des couplages mécanique-chimie-transfert dans la tenue à long terme des ouvrages de stockage de déchets radioactifs, PhD Thesis, E.N.S. de Cachan and University of Laval, 1996.

[GÉR 98] GÉRARD B., PIJAUDIER-CABOT G., LA BORDERIE C., "Coupled diffusion-damage modelling and the implications on failure due to strain localisation", *International Journal of Solids and Structures,* vol. 35, pp. 4107–4120, 1998.

[GUR 77] GURSON A.L., "Continuum theory of ductile rupture by void nucleation and growth: Part I-yield criterion and flow rules for porous ductile media", *Transactions of the ASME,* vol. 99, pp. 2–15, 1977.

[HEU 01] HEUKAMP F., ULM F.-J., GERMAINE J., "Mechanical properties of calcium leached cement paste: Triaxial stress states and the influence of pore pressure", *Cement and Concrete Research,* vol. 31, pp. 767–774, 2001.

[KUH 04] KUHL D., BANGERT F., MESCHKE G., "Coupled chemo-mechanical deterioration of cementitious materials, Part I: Modeling", *International Journal of Solids and Structures*, vol. 41, pp. 15–40, 2004.

[MOR 73] MORI T., TANAKA K., "Average stress in matrix and average elastic energy of materials with misfitting inclusions", *Acta Metallurgica,* vol. 21, pp. 571–574, 1973.

[PIE 88] PIETRUSZCZAK S., JIANG J., MIRZA F.A., "An elastoplastic constitutive model for concrete", *International Journal of Solids and Structures,* vol. 24, no. 7, pp. 705–722, 1988.

[XIE 06] XIE S.Y., SHAO J.F., BURLION N., "Experimental study of mechanical behaviour of cement paste under compressive stress and chemical degradation", *Cement and Concrete Research,* vol. 38, no. 12, pp. 1416–1423, 2008.

[YUR 11a] YURTDAS I., XIE S.Y., BURLION N., SHAO J.F., SAINT-MARC J., GARNIER A., "Influence of chemical degradation on mechanical behaviour of a petroleum cement paste", *Cement and Concrete Research,* vol. 41, pp. 412–421, 2011a.

[YUR 11b] YURTDAS I., XIE S.Y., BURLION N., SHAO J.F., SAINT-MARC J., GARNIER A., "Deformation and permeability evolution of petroleum cement paste subjected to chemical degradation under temperature", *Transport in Porous Media,* vol. 86, no. 3, pp. 719–736, 2011b.

Chapter 10

Reactive Transport Modeling of CO_2 Through Cementitious Materials Under Supercritical Boundary Conditions

10.1. Introduction

Geological storage of CO_2 is considered as a promising solution to global warming arising from anthropogenic CO_2 emissions. However, long-term storage of CO_2 involves risks of leakage of CO_2 through the oil well cement-based material. To predict CO_2 leakage, it is essential to understand CO_2 transfers, as well as the chemical reactions at stake (carbonation mechanism), which take place within the materials exposed to CO_2 under geological storage conditions.

In deep geological formations (more than 1 km deep), the temperature could reach 50 °C or even more, and the pressure could be more than 10 MPa. Under these conditions, CO_2 is in a supercritical state ($scCO_2$), with the density like that of a liquid and the viscosity like a gas. Since $scCO_2$ and water are not miscible and can dissolve into each other, they form a two-phase system composed of an $scCO_2$-rich phase (wet $scCO_2$) and a water-rich phase (CO_2-saturated aqueous solution). Therefore, the material could be exposed to either a CO_2-saturated aqueous solution or wet $scCO_2$, or even both.

Research work conducted in this field includes experimental studies as well as numerical modeling.

Chapter written by Jitun SHEN, Patrick DANGLA and Mickaël THIERY.

Duguid and Scherer [DUG 10] conducted a series of experiments to examine the effects of flowing carbonated brine on well cementitious materials. They used CO_2 saturated brine in a batch reactor and simulated both sandstone and limestone reservoir conditions (i.e. boundary conditions saturated with calcite or not). Class H cement pastes were exposed to different test conditions simulating sandstone-like conditions: $T = 20$ or $50\,°C$ and $pH = 2.4$ or 3.7. All the samples under sandstone-like conditions were degraded over the course of the experiment.

Under limestone-like conditions, no evident attack was observed. Duguid [DUG 09] also made some proposals to predict the *in situ* degradation of an oil-well cement in contact with sandstone. The authors believe that the formation of a calcite-rich layer will largely hinder carbonation penetration. Thus, with a rough estimation, it will take 30,000–70,000 years to degrade 25 mm of a cement paste in a sandstone reservoir.

Rimmele *et al.* [RIM 08] exposed cement samples to both CO_2-saturated brine and wet $scCO_2$, under pressure and temperature similar to downhole conditions. The porosity distribution at different ages was monitored by SEM-BSE image analysis. The porosity evolution showed an initial sealing stage related to calcium carbonation precipitation clogging the porosity, followed by a dissolution stage, which is more efficient in CO_2-saturated water condition than in wet $scCO_2$ condition.

Existing models mainly focus on the transport and reactions within cement-based materials exposed to CO_2-saturated brine. Carey *et al.* [CAR 07] have presented numerical simulations that are consistent with *in situ* observations if the modeling parameters are adjusted. An interesting reactive transport modeling has been proposed and discussed in detail in the work of Huet *et al.* [HUE 10]. The authors have intended to simulate the conditions of CO_2-saturated water, which were tested in Duguid's experiments [DUG 10]. Despite these simulation works, comparisons between experimental works and numerical simulations are far from being abundant in the literature. Indeed published modeling works mainly deal with saturated conditions, that is CO_2-saturated brine. Since in the case of CO_2 geological storage conditions, cement-based materials could also be exposed to $scCO_2$, a model should work with both conditions, wet $scCO_2$ and CO_2-saturated water conditions.

This work proposes a reactive transport modeling of CO_2 throughout cement paste. The cement can be exposed to both CO_2-saturated brine and wet $scCO_2$. The chemistry of the CO_2-cement system is simplified to the dissociation reactions of the three main minerals: portlandite (CH), calcium silicate hydrates (C-S-H) and calcite ($C\bar{C}$). Basically, the reactivity of the cement to CO_2 results in the dissolution of portlandite (CH) and calcium silicate hydrates (C-S-H) as well as the precipitation of calcite ($C\bar{C}$) and amorphous silica (SH_t). Completely dissolved mineral can be accounted for through inequalities involving ion activity products of the dissociation reactions of minerals. A thermodynamic description of the dissolution of C-S-H is proposed to account for the continuous variation (decrease) of the C/S ratio during the dissolution reaction of C-S-H. Amorphous silica (SH_t) results here from a completely decalcified state of C-S-H.

The key parameters implemented in this numerical code are detailed and discussed in the following sections. Simulations are conducted, including saturated condition of Duguid's experiments [DUG 10] and unsaturated conditions presented in Rimmele's work [RIM 08]. Some simulation results dealing with carbonation and porosity profiles are pointed out and compared with experimental results taken from the previous references. With the help of modeling, it is possible to provide some explanations for the observed phenomena.

10.2. Carbonation of cement-based materials

The carbonation of cement-based materials is a chemical reaction between the dissolved CO_2 in the aqueous pore solution and the hydration products. Therefore, the solubility of the supercritical CO_2 gas is required first.

10.2.1. *Solubility of the supercritical CO_2 in the pore solution*

We have followed the work of Spycher *et al.* [SPY 03] to calculate the CO_2 solubility in water in the range of temperature 12–110°C and at gas pressure up to 600 bars. The gas phase is a binary mixture of CO_2 and H_2O. The mole fraction of H_2O is denoted with $y_{\mathrm{H_2O}}$. The model proposed by Spycher *et al.* can reproduce fugacity coefficients of CO_2 and H_2O within 1% of published reference data. The CO_2 solubility, $s_{\mathrm{CO_2}}$, and the mole fraction of water in the gas phase, $y_{\mathrm{H_2O}}$, are given in terms of the gas and liquid pressure after Spycher *et al.* [SPY 03] as follows:

$$s_{\mathrm{CO_2}} = \frac{\phi_{\mathrm{CO_2}}(P_{\mathrm{G}})}{K_{\mathrm{CO_2}}(P_{\mathrm{L}})} P_{\mathrm{G}}(1 - y_{\mathrm{H_2O}}) \qquad [10.1]$$

$$y_{\mathrm{H_2O}} = \frac{K_{\mathrm{H_2O}}(P_{\mathrm{L}})}{\phi_{\mathrm{H_2O}}(P_{\mathrm{G}})P_{\mathrm{G}}} a_{\mathrm{H_2O}} \qquad [10.2]$$

where $a_{\mathrm{H_2O}}$ is the chemical activity of water in the pore solution assumed equal to unity here. In equations [10.1] and [10.2], P_{G} and P_{L} are the gas and liquid pressures while $\phi_{\mathrm{CO_2}}$ and $\phi_{\mathrm{H_2O}}$ are the fugacity coefficients as obtained by Spycher. Moreover, $K_{\mathrm{CO_2}}(P_{\mathrm{L}})$ and $K_{\mathrm{H_2O}}(P_{\mathrm{L}})$ are given by:

$$K_{\mathrm{CO_2}}(P_{\mathrm{L}}) = K^0_{\mathrm{CO_2}} \exp\left(\frac{(P_{\mathrm{L}} - P^{\ominus})V_{\mathrm{CO_2}}}{RT}\right) \qquad [10.3]$$

$$K_{\mathrm{H_2O}}(P_{\mathrm{L}}) = K^0_{\mathrm{H_2O}} \exp\left(\frac{(P_{\mathrm{L}} - P^{\ominus})V_{\mathrm{H_2O}}}{RT}\right) \qquad [10.4]$$

V_i is the partial molar volume of i in the standard state. Here, we use $V_{H_2O} = 18.1$ cm^3/mol, $V_{CO_2} = 32.6$ cm^3/mol as in the work of Spycher *et al.* [SPY 03]. $P^{\ominus}$ is a standard reference pressure taken as 1 bar.

The fugacity coefficients ϕ_i are reproduced in Figure 10.1.

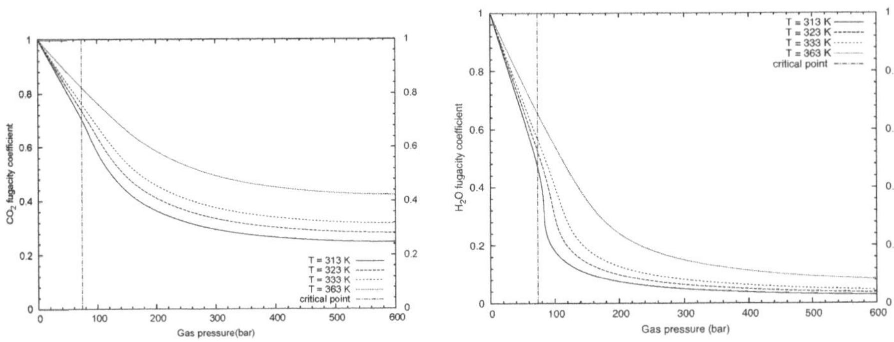

Figure 10.1. *Fugacity coefficients of CO_2 and H_2O at different temperature after Spycher et al. [SPY 03]*

10.2.2. *Chemical reactions*

Therefore the dissolved CO_2, denoted as CO_2^0 hereafter, will acidify the pore solution and react with calcium-based and silicon-based ions resulting in a series of homogeneous chemical reactions as shown in Table 10.1. Mass action laws are considered for these reactions. Note that, in addition to these ions, alkalis K^+/Na^+ and chloride Cl^- are included in the model. The effects of the chemical activity are disregarded as a first approximation (a unit activity coefficient is assumed throughout this Chapter).

We assume that only the portlandite (CH) and the calcium silicate hydrates (C-S-H), which are the main hydrates of cement, react with the dissolved CO_2 to form calcium carbonate ($C\bar{C}$)[1]. These global reactions can be understood by combining three basic dissociation reactions taking place between these minerals and the aqueous pore solution in the form:

$$CH \rightleftharpoons Ca^{2+} + 2OH^- \quad [10.5]$$

$$C\bar{C} \rightleftharpoons Ca^{2+} + CO_3^{2-} \quad [10.6]$$

$$C_xS_yH_z \rightleftharpoons xCa^{2+} + 2xOH^- + ySiO_2^0 + (z-x)H_2O \quad [10.7]$$

1 We will use the cement chemistry notation throughout the paper: C = CaO, S = SiO_2, H = H_2O, $\bar{C}$ = CO_2.

Aqueous reactions			log K
H_2O	$\rightleftharpoons$	$H^+ + OH^-$	-14
$CO_2^0 + H_2O$	$\rightleftharpoons$	H_2CO_3	-2.77
H_2CO_3	$\rightleftharpoons$	$HCO_3^- + H^+$	-3.6
HCO_3^-	$\rightleftharpoons$	$CO_3^{2-} + H^+$	-10.25
$Ca(OH)_2^0$	$\rightleftharpoons$	$Ca^{2+} + 2OH^-$	0
$CaCO_3^0$	$\rightleftharpoons$	$Ca^{2+} + CO_3^{2-}$	-3.145
$CaOH^+$	$\rightleftharpoons$	$Ca^{2+} + OH^-$	-1.22
$CaHCO_3^+$	$\rightleftharpoons$	$Ca^{2+} + HCO_3^-$	-1.106
$CaH_2SiO_4^0$	$\rightleftharpoons$	$Ca^{2+} + H_2SiO_4^{2-}$	-4.6
$CaH_3SiO_4^-$	$\rightleftharpoons$	$Ca^{2+} + H_3SiO_4^-$	-1.2
$SiO_2^0 + 2H_2O$	$\rightleftharpoons$	$H_4SiO_4^0$	-2.71
$H_4SiO_4^0$	$\rightleftharpoons$	$H_3SiO_4^- + H^+$	-9.81
$H_3SiO_4^-$	$\rightleftharpoons$	$H_2SiO_4^{2-} + H^+$	-13.33

Table 10.1. *Chemical reactions in the aqueous solution. Equilibrium constants are provided for standard conditions [THO 03]*

These basic dissociation reactions can be considered either in thermodynamical equilibrium or not, depending on the concentration of the dissolved CO_2. First let us consider the carbonation of CH. That of C-S-H will be considered in the following section since a new approach will be proposed to account for the continuous change of the stoichiometric coefficients of C-S-H.

10.2.3. *Carbonation of CH*

We can infer from the thermodynamical analysis of reactions [10.5] and [10.6] that the stability of the portlandite and the calcium carbonate can be characterized by a threshold of the ion activity products in the form:

$$Q_{\mathrm{CH}} = a_{\mathrm{Ca}^{2+}} a_{\mathrm{OH}^-}^2 \leq K_{\mathrm{CH}} \qquad [10.8]$$

$$Q_{\mathrm{C\bar{C}}} = a_{\mathrm{Ca}^{2+}} a_{\mathrm{CO}_3^{2-}} \leq K_{\mathrm{C\bar{C}}} \qquad [10.9]$$

where $pK_{\mathrm{CH}} = 5.2$ and $pK_{\mathrm{C\bar{C}}} = 8.36$. Using the chemical equilibriums listed in Table 10.1, we easily derive:

$$\frac{Q_{\mathrm{C\bar{C}}}}{K_{\mathrm{C\bar{C}}}} = \frac{Q_{\mathrm{CH}}}{K_{\mathrm{CH}}} \frac{\rho_{\mathrm{CO}_2^0}}{\rho_{\mathrm{CO}_2^0}^{\mathrm{CH}}} \qquad [10.10]$$

where $\rho_{CO_2^0}$ is the dissolved CO_2 concentration (i.e. $\rho_{CO_2^0} = s_{CO_2}$ in the presence of supercritical gas) and $\rho^{CH}_{CO_2^0} \approx 3.10^{-15}$ mol/L calculated with the help of the constants listed in Table 10.1. Therefore, $C\bar{C}$ is not stable for $\rho_{CO_2^0} < \rho^{CH}_{CO_2^0}$ (since $Q_{C\bar{C}}$ is necessarily lower than $K_{C\bar{C}}$) and CH is not stable for $\rho_{CO_2^0} > \rho^{CH}_{CO_2^0}$. These stability domains are summarized in Figure 10.2.

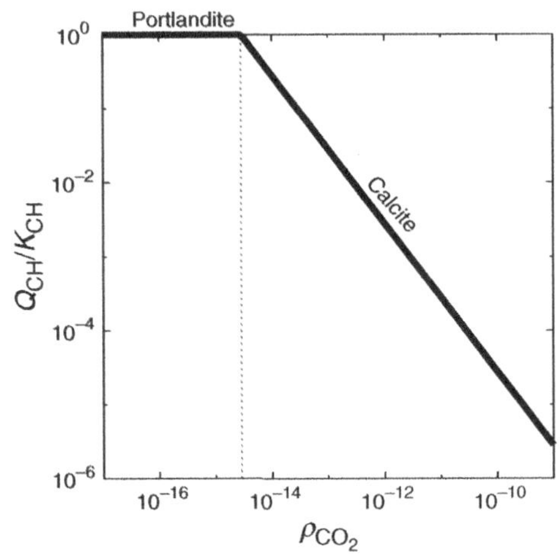

Figure 10.2. *Stability domains of portlandite and calcium carbonate*

Before carbonation, that is for $\rho_{CO_2^0} < \rho^{CH}_{CO_2^0}$, portlandite is stable. When CO_2 concentration gets higher and exceeds $\rho^{CH}_{CO_2^0}$, portlandite is not stable anymore and starts to dissolve. A kinetic law for this dissolution process is introduced to at least facilitate numerical convergence. Therefore, a simple law for the rate of decrease of the CH mole content can be formulated by using a characteristic time, τ_{CH}, as follows:

$$\frac{dn_{CH}}{dt} = -\frac{n_{CH}}{\tau_{CH}} \ln\left(\frac{\rho_{CO_2^0}}{\rho^{CH}_{CO_2^0}}\right) \qquad [10.11]$$

Owing to the high CO_2 concentration used in this chapter, this characteristic time is chosen in practice as small as possible to oblige the reaction to be as close as possible to equilibrium. The dissolved calcium ion will precipitate into $C\bar{C}$ with the almost same rate as that of the portlandite dissolution. For lower CO_2 concentration such as found in atmospheric carbonation, a more physical kinetic law, as found in Thiery *et al.* [THI 06], should be used.

Finally, this model intends to account for a possible dissolution of the calcium carbonate $C\bar{C}$ when the pH falls down to a small enough value. For this, we define the following variable as the main unknown of the numerical procedure:

$$\zeta_C = \frac{n_{C\bar{C}}}{n^0} + \log\left(\frac{Q_{C\bar{C}}}{K_{C\bar{C}}}\right) \quad \text{for} \quad \rho_{CO_2^0} > \rho_{CO_2^0}^{CH} \tag{10.12}$$

$$\zeta_C = \frac{n_{CH}}{n^0} + \log\left(\frac{Q_{CH}}{K_{CH}}\right) \quad \text{for} \quad \rho_{CO_2^0} < \rho_{CO_2^0}^{CH} \tag{10.13}$$

This method enables us to capture the precipitation and dissolution of the relevant solid phase. A positive value for ζ_C stands for the molar content of the precipitated phase while a negative value stands for the ion activity product of its dissociation reaction.

10.2.4. *Carbonation of C-S-H*

The carbonation of C-S-H results from the two basic dissociation reactions [10.6] and [10.7]. Therefore, modeling the dissolution of the C-S-H [10.7] is an important requirement to understand the process of carbonation of cement. To do so, we have to model the thermodynamic properties of C-S-H including its incongruent behavior. Literature provides serveral kinds of modeling, starting from the use of empirical or semi-empirical models [REA 90, REA 92] and evolving to solid solution models [KUL 01]. We propose here a more general description of the solubility that encompasses the solid solution model. It is known that the C/S ratio of C-S-H is variable. C/S ratio is about 1.7 from fresh hydrated Portland cement and tends to lower during a dissolution process due to leaching or carbonation. The C/S ratio is therefore considered as a variable of the model. Defining a unit mole of C-S-H as a unit mole of Si (i.e. $y = 1$), the dissociation reaction of an infinitesimal small amount of C-S-H (to consider x and z as constants) can be written formally as:

$$C_xS_1H_z \rightleftharpoons xCa^{2+} + 2xOH^- + SiO_2^0 + (z - x)H_2O \tag{10.14}$$

where x stands for the current C/S ratio and z for the H/S ratio. As defined implicitly, molar quantities related to C-S-H are defined, in the following, per unit mole of Si. Accordingly, the molar Gibbs energy of C-S-H, $g(x, z)$, depends only on x and z (in addition to the pressure and the temperature that are not explicitly mentionned for sake of simplicity). During the dissociation of an infinitesimally small amount of moles of C-S-H, dn, the Gibbs free energy of [10.14], G, can only decrease spontaneously:

$$\begin{aligned} dG = &-\mu_{Ca^{2+}}d(nx) - \mu_{OH^-}d(2nx) - \mu_{SiO_2^0}dn - \mu_{H_2O}d(n(z - x)) \\ &+d(ng(x, z)) \leq 0 \end{aligned} \tag{10.15}$$

where μ_i are the chemical potentials. We will assume, later on, that z is a function of x: $z(x)$. Therefore, the molar Gibbs energy of C-S-H is $g(x)$. At equilibrium, G is minimum and thereby, noting $z' = dz/dx$, we get:

$$\left(-x(\mu_{\mathrm{Ca}^{2+}} + 2\mu_{\mathrm{OH}^-}) - \mu_{\mathrm{SiO}_2^0} - (z - x)\mu_{\mathrm{H_2O}} + g(x)\right) dn + \left(-\mu_{\mathrm{Ca}^{2+}} - 2\mu_{\mathrm{OH}^-} - (z' - 1)\mu_{\mathrm{H_2O}} + \frac{\partial g}{\partial x}\right) n dx = 0 \quad [10.16]$$

for arbitrary dn and n. Since the chemical potentials cannot depend on n, we must conclude that:

$$\mathcal{A} = x(\mu_{\mathrm{Ca}^{2+}} + 2\mu_{\mathrm{OH}^-}) + \mu_{\mathrm{SiO}_2^0} + (z - x)\mu_{\mathrm{H_2O}} - g(x) = 0 \quad [10.17]$$

$$\frac{\partial \mathcal{A}}{\partial x} = (\mu_{\mathrm{Ca}^{2+}} + 2\mu_{\mathrm{OH}^-}) + (z' - 1)\mu_{\mathrm{H_2O}} - \frac{\partial g}{\partial x} = 0 \quad [10.18]$$

Equation [10.17] states that the chemical affinity of reaction [10.14] vanishes at equilibrium. Therefore, this equation will provide an equation in the form of [10.21] similar to the mass action law. Equation [10.18] states that this chemical affinity must remain at zero or equivalently that equilibrium must hold during the decalcification of the C-S-H. This equation will provide a new equation in the form of [10.22]. Indeed, classical expression for the chemical potential entails:

$$\mu_{\mathrm{Ca}^{2+}} + 2\mu_{\mathrm{OH}^-} = \mu^0_{\mathrm{CH}} + RT \ln\left(\frac{Q_{\mathrm{CH}}}{K_{\mathrm{CH}}}\right) \qquad (Q_{\mathrm{CH}} = a_{\mathrm{Ca}^{2+}} a^2_{\mathrm{OH}^-}) \quad [10.19]$$

$$\mu_{\mathrm{SiO}_2^0} + t\mu_{\mathrm{H_2O}} = \mu^0_{\mathrm{SH}_t} + RT \ln\left(\frac{Q_{\mathrm{SH}_t}}{K_{\mathrm{SH}_t}}\right) \qquad (Q_{\mathrm{SH}_t} = a_{\mathrm{SiO}_2^0} a^t_{\mathrm{H_2O}}) \quad [10.20]$$

where $t = z(0)$ is the hydration level of the amorphous silica, SH_t, obtained after a complete decalcification of the C-S-H. K_{CH} and K_{SH_t} are the equilibrium constants for the dissolution of portlandite and amorphous silica. From the above equations we have:

$$x \ln\left(\frac{Q_{\mathrm{CH}}}{K_{\mathrm{CH}}}\right) + \ln\left(\frac{Q_{\mathrm{SH}_t}}{K_{\mathrm{SH}_t}}\right) = \ln a(x) \quad [10.21]$$

$$\ln\left(\frac{Q_{\mathrm{CH}}}{K_{\mathrm{CH}}}\right) = \frac{\partial \ln a}{\partial x} \quad [10.22]$$

where we defined $RT \ln a(x) = g(x) - x\mu^0_{\text{CH}} - \mu^0_{\text{SH}_t} - (z-t)\mu_{\text{H}_2\text{O}}$. Note that $\ln a(0) = 0$ since we have to meet the molar Gibbs energy of the amorphous silica in the form $g(0) = \mu^0_{\text{SH}_t}$. Equation [10.22] shows that x and $\frac{Q_{\text{CH}}}{K_{\text{CH}}}$ are related through a constitutive-like equation of the form:

$$x = \chi\left(\frac{Q_{\text{CH}}}{K_{\text{CH}}}\right) \qquad [10.23]$$

From [10.21] and [10.23], we infer that there is a relationship between $Q_{\text{SH}_t}/K_{\text{SH}_t}$ and $Q_{\text{CH}}/K_{\text{CH}}$ in the form:

$$\ln \frac{Q_{\text{SH}_t}}{K_{\text{SH}_t}} = -\int_0^{\frac{Q_{\text{CH}}}{K_{\text{CH}}}} \frac{\chi(q)}{q} dq \qquad [10.24]$$

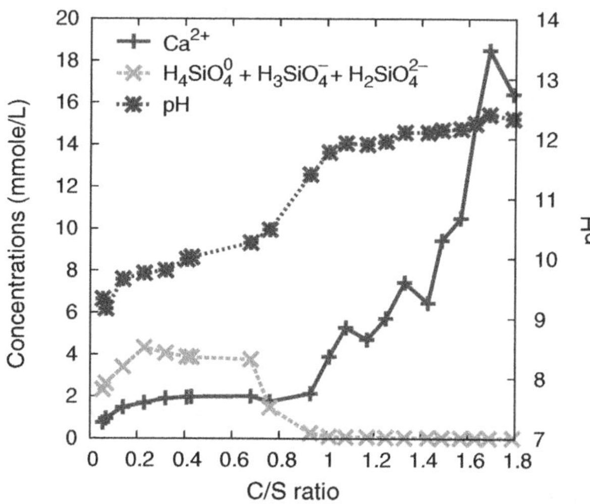

Figure 10.3. *Composition of a calcium–silicon solution in equilibrium with its solid phase versus C/S ratio. Results reported from Greenberg and Chang [GRE 65]*

Equation [10.24] is a generalization of the mass action law for C-S-H with variable C/S ratio. To confirm this relationship, we used the result of an experiment conducted by Greenberg and Chang [GRE 65]. In these experiments, the solubilities of reaction mixtures of calcium oxide, silica and water were investigated (0.805 g of SiO_2 was poured in 100 mL of water with varying additions of CaO). The calcium ion concentration, the silicic acid concentration (H_4SiO_4 + $H_3SiO_4^-$ + $H_2SiO_4^{2-}$), and the pH were measured. They are reported in Figure 10.3. In Figure 10.4, we reported the C/S ratio versus Q_{CH}/K_{CH} obtained by Greenberg and Chang and where the ion activity product, Q_{CH}, was calculated with the ion concentrations, thereby disregarding the chemical activity effects. We get a smooth curve as suggested by relation [10.23]. We also plotted the fraction Q_{SH_t}/K_{SH_t} obtained on one hand from equation [10.24], and on the other hand from the direct use of the Greenberg and Chang's results. To perform this latter calculation, the ion activity product of silica gel, Q_{SH_t}, is approximated by the concentration of the monosilicic acid $H_4SiO_4^0$. The solubility constant, K_{SH_t}, is 1.93 mM. The comparison between the model [10.24] and the experiment is fairly good.

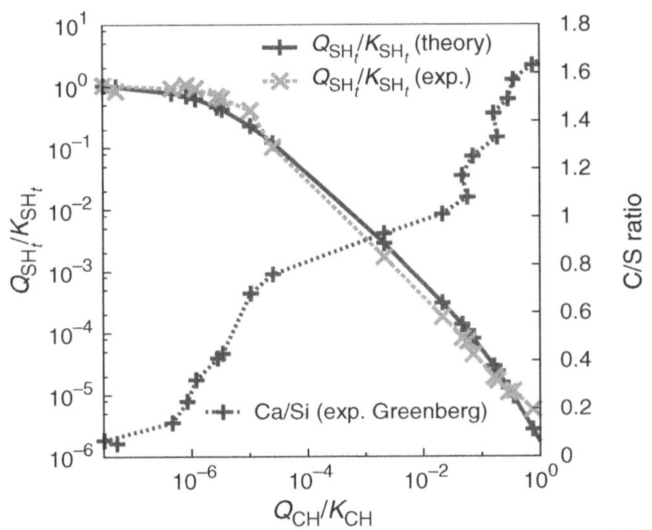

Figure 10.4. *The fraction Q_{SH_t}/K_{SH_t} computed with equation* [10.24] *is compared to the direct use of the Greenberg's and Chang's experiment*

10.2.5. *Porosity change*

The change in porosity, $\phi - \phi_0$, induced by the precipitation-dissolution of the various solid compounds can easily be accounted for by a simple balance of volume:

$$\phi - \phi_0 = -V_{CH}(n_{CH} - n_{CH}^0) - V_{C\bar{C}}(n_{C\bar{C}} - n_{C\bar{C}}^0) - V_{C-S-H}n_{Si} + V_{C-S-H}^0 n_{Si}^0 \quad [10.25]$$

In equation [10.25], V_i is the molar volume of the solid compound i (L/mol), n_i and n_i^0 are the current and initial solid content of i (mol/L). We used $V_{CH} = 33$ cm^3/mol for the molar volume of portlandite and $V_{C\bar{C}} = 37$ cm^3/mol for the molar volume of calcite.

However, there is a lack of data regarding the molar volume of C-S-H. It is theoretically a function of the C/S ratio, x, and the H/S ratio, z, as seen previously (V_{C-S-H} is the partial derivative of g with respect to the pressure). The molar volume of $C_{1.7}SH_{1.4}$ is approximately 64.5 cm^3/mol, that of $C_{1.7}SH_{2.1}$ is 78.8 cm^3/mol, and that of $C_{1.7}SH_4$ is 113.6 cm^3/mol [JEN 04]. On the other hand, that of amorphous silica (unhydrated i.e. $t = 0$) is approximately 29 cm^3/mol [LOT 08]. Therefore, in the absence of knowledge we use a linear function for $V_{C-S-H}(x) : V_{C-S-H}(x) = x/x_0 V^0_{C-S-H} + (1 - x/x_0)V_{SH_t}$ with $x_0 = 1.7$, $V_{SH_t} = 29$ cm^3/mol and $V^0_{C-S-H} = 78$ cm^3/mol. Consistently, we choose an H/S ratio as $z(x) = x/x_0 z_0 + (1 - x/x_0)t$ with $t = 0$ and $z_0 = 2$.

10.3. Reactive transport modeling

The model aims at simulating the behavior of a water-saturated cement paste in contact with a CO_2-saturated brine or a supercritical CO_2 gas. The major solid components including calcium hydroxide (CH) and calcium silicate hydrates (C-S-H) are initially assessed (by direct measurements or by using simple analytical models). When put in contact with the cement, the CO_2 will diffuse through the pores of the cement paste. When the CO_2 concentration gets high enough ($>3 \times 10^{-15}$ mol/L), CH will start to dissolve and calcium carbonate will form. When CO_2 concentration gets higher, C-S-H carbonates and generates calcium carbonate. During this carbonation process, the microstructure of cement also changes. When portlandite dissolves, the porosity will first increase, and it will then decrease as the calcite forms. The decalcification of C-S-H could also contribute to the porosity increase or drop, due to the volume difference between C-S-H and the precipitated calcium carbonate [THI 12]. When the pH value of the system drops below five (or even lower), the formed calcite could dissolve under peculiar boundary conditions and cause an increase in porosity. The transport of calcium from the carbonated region to the non-carbonated one could also be responsible for a porosity increase between the carbonation fronts of CH and C-S-H. All these phenomena are included in the modeling presented below.

10.3.1. *Field equations*

The coupling between the transport and the chemical reactions is treated thanks to a set of mass balance equations. The first subset of equations consists of mass balance equations applied to atoms such as carbon (C), calcium (Ca), silicon (Si), potassium (K), and chlorine (Cl). For such atom (A = C, Ca, Si, K, Cl) we have:

$$\frac{\partial n_A}{\partial t} = -\text{div } \underline{w_A} \qquad [10.26]$$

where n_A stands for the total molar content of atom A per unit volume of the porous medium. There are three contributions to n_A associated to the liquid, gas and solid phases: $n_A = \phi S_L \rho_A^L + \phi S_G \rho_A^G + n_A^S$. For example, let us consider the carbon element A = C. The carbon concentration in the liquid phase is:

$$\rho_C^L = \rho_{CO_2^0} + \rho_{H_2CO_3} + \rho_{HCO_3^-} + \rho_{CO_3^{2-}} + \rho_{CaHCO_3^+} + \rho_{CaCO_3^0} \qquad [10.27]$$

where the ρ_i are linked by mass action laws and equilibrium constants listed in Table 10.1. The concentration in the gas phase is:

$$\rho_C^G = (1 - y_{H_2O})V_G \qquad [10.28]$$

where y_{H_2O} is given by [10.2] and V_G is the molar volume of the gas mixture. Following the work of Spycher *et al.* [SPY 03]), the Redlich–Kwong's model [RED 49] was used for the equation of state (EOS) of the gas. As proposed by Spycher, the influence of water is neglected in this EOS so it is the same as that of pure CO_2. The compression factor $Z(P_G) = P_G V_G / RT$ of the Redlich–Kwong's model is plotted in Figure 10.5. Finally, the carbon solid content is the mole content of the calcium carbonate: $n_C^S = n_{C\bar{C}}$.

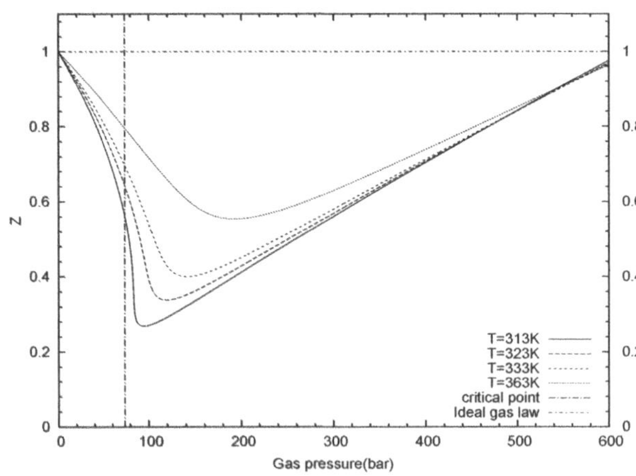

Figure 10.5. *Compression factor, Z, of scCO$_2$ at different temperature, after Nicolas Spycher et al. [SPY 03]*

Obviously the total molar flow of A, $\underline{w_A}$, can be decomposed in the same manner as n_A. The transport of species will be given in the section below.

The second subset of equations to be solved could have been mass balance equations for the hydrogen (H) and oxygen (O) atoms. Instead, it is more convenient to consider the balance equations for the total mass on the one hand and the electric charge carried by the ions on the other hand. The first one reads

$$\frac{\partial m}{\partial t} = -\text{div}\ \underline{w} \qquad [10.29]$$

In equation [10.29] m stands for the total mass. There are three contributions to m: $m = \phi S_L \rho_L + \phi S_G \rho_G + m_S$. Similarly there are two contributions to the total mass flow, $\underline{w} : \underline{w} = \rho_L \underline{v_L} + \rho_G \underline{v_G}$. These liquid and gas mass flows are given by a generalized Darcy's law and recalled in the sections below.

The electric charge balance equation results from the fact that each molecule, i, carries a constant charge proportional to its valence number, z_i, and that each chemical reaction does not provide any source of electric charge (there is a balance of charge for each chemical reaction). As a result, there is a global electric charge balance equation of the form:

$$\text{div}\ \underline{i} = 0 \qquad [10.30]$$

where $\underline{i}$ stands for the ionic current reading

$$\underline{i} = \sum_i z_i \underline{w_i} \qquad [10.31]$$

where the summation applies on the set of electrolyte ions. The ionic flow $\underline{w_i}$ is given by a Nernst–Planck equation [10.36]. In this law, each ion has its own diffusion coefficient according to its size. Therefore, an electric potential must be generated in the medium to provide at any time an electroneutral pore solution. As a result, electroneutrality must hold in the medium in the form:

$$\sum_i z_i \rho_i = 0 \qquad [10.32]$$

Equations [10.26–10.32] are the set of field equations governing the coupling between transport and chemistry. In these equations, the saturation degree S_L ($S_G = 1 - S_L$) is a function of the capillary pressure P_c, difference between gas and liquid pressures: $P_c = P_G - P_L$. This function strongly depends on the cement microstructure, as well as the temperature. An example of the relationship between S_L and P_c is shown in Figure 10.6.

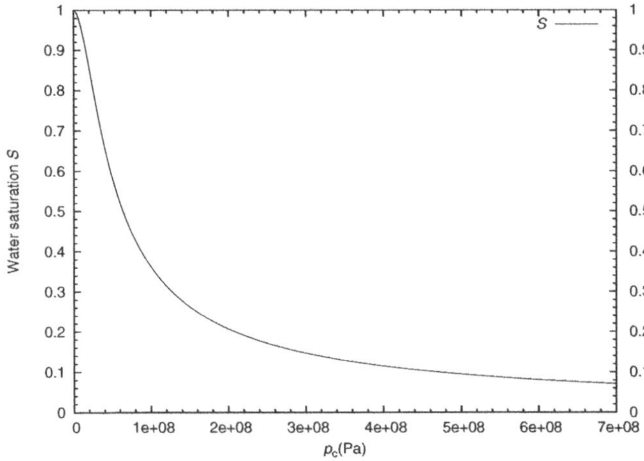

Figure 10.6. *Typical relationship between the liquid saturation S_L and the capillary pressure P_c (case of an OPC paste, w/c = 0.45 [BAR 07])*

10.3.2. *Transport of the liquid phase*

For the liquid transport, we consider the advection of pore solution due to liquid pressure gradient. Darcy's law is employed here, written as follows:

$$\underline{v_{\mathrm{L}}} = -\frac{K}{\mu_{\mathrm{L}}} k_{rl}(S_{\mathrm{L}}) \underline{\nabla P_{\mathrm{L}}} \qquad [10.33]$$

where μ_{L} is the dynamic viscosity of liquid, K is the intrinsic permeability of porous material. Here, we consider K as a function of porosity, following the work of van Genuchten [VAN 80]:

$$K = K^0 \left(\frac{\phi}{\phi_0}\right)^3 \left(\frac{1-\phi_0}{1-\phi}\right)^2 \qquad [10.34]$$

where K^0 is the intrinsic permeability before carbonation, with the porosity of ϕ_0. The relative permeability to liquid, $k_{rl}(S_{\mathrm{L}})$, can be generated from the $P_c - S_{\mathrm{L}}$ curve following the work of Mualem [MUA 76]. Figure 10.7 shows an example of relative permeability obtained for a cement paste for which the liquid phase disconnects below a saturation degree of 0.36.

10.3.3. *Transport of the gas phase*

The transport of gas is similar to the liquid transport. The transport equation is written as:

$$\underline{v_{\mathrm{G}}} = -\frac{K}{\mu_{\mathrm{G}}} k_{rg}(S_{\mathrm{L}}) \underline{\nabla P_{\mathrm{G}}} \qquad [10.35]$$

where μ_G is the dynamic viscosity of the gas, which strongly depends on the gas pressure and temperature, according to the work of Fenghour *et al.* [FEN 98], (see Figure 10.8). K is the intrinsic permeability of cement as seen previously while $k_{rg}(S_L)$ is the relative permeability to gas, which can be generated in the same way as for liquid phase according to the Mualem's model (see Figure 10.9).

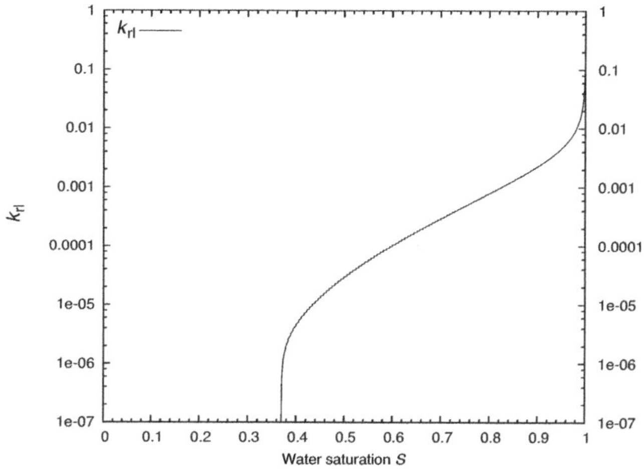

Figure 10.7. *A typical curve for the relative permeability to liquid (case of OPC paste with a w/c = 0.45)*

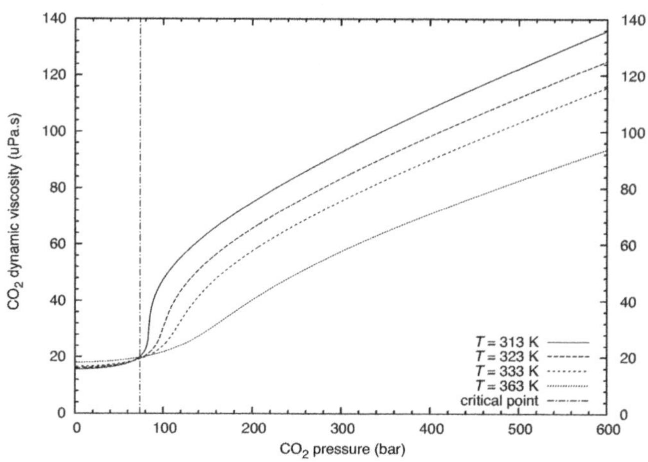

Figure 10.8. *CO_2 viscosity at different temperature after Fenghour et al. [FEN 98]*

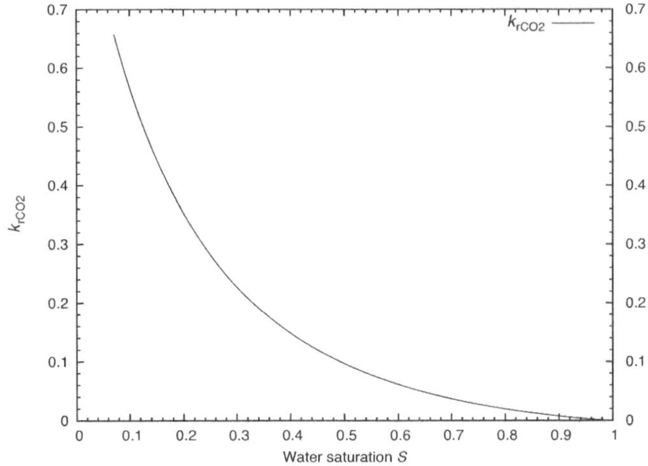

Figure 10.9. *A typical curve for the relative permeability to gas (case of an OPC paste, w/c= 0.45)*

10.3.4. *Transport of aqueous species*

The transport of aqueous species is governed by the electro-diffusion (Nernst–Plank equation) and the advection, as indicated in equation [10.36]. In the electro-diffusion term, F is the Faraday's constant, R is the perfect gas constant, T is the absolute temperature while z_i and ρ_i stand for the ionic valence and the concentration. The first part refers to the diffusion effect due to the concentration gradient (in agreement with the Fick's law), and the second part accounts for the electric effect. In the advection term, $\underline{v_L}$ is the velocity of the liquid phase as discussed in section 10.3.2.

$$\underline{w_i} = -D_i \left(\underline{\nabla \rho_i} + \rho_i \frac{F z_i}{RT} \underline{\nabla \psi} \right) + \rho_i \underline{v_L} \qquad [10.36]$$

The effective diffusion coefficient D_i of each species i is calculated by taking into account porosity changes and the saturation degree [TOG 98].

$$D_i(\phi, S_L) = D_i^0 \, 2.9 \, 10^{-4} e^{9.95\phi} S_L^{\lambda} \qquad [10.37]$$

with D_i^0 as the diffusion coefficient in pure water and $\lambda = 4$ according to Baroghel–Bouny [BAR 07]. Note that the diffusion coefficients of Na^+ and K^+ are quite similar. It is thus acceptable to assimilate these two species.

10.4. Simulation results and discussion

Some calculations are conducted to simulate experimental works taken from Duguid and Scherer [DUG 10]. In this experiment, the author exposed a cylindrical class of H cement samples (7.5 mm in diameter) to a CO_2 saturated brine in atmospheric conditions (20°C, 1 atm, pure bubbling of CO_2) for 30 days in a batch reactor. In one set of experiments, the brine is saturated with calcite before putting in contact with the cement paste, thus simulating limestone-like reservoir conditions, in the other set the brine is not saturated with calcite standing for sandstone-like reservoir conditions. To simulate Duguid's experiments, we consider a cement paste cylinder with the same diameter, divided into 50 nodes. The initial CO_2 concentration in the pore solution of the material is small enough to avoid the dissolution of the portlandite (here equal to 10^{-15} mol/L), while at the boundary this concentration is 0.057 mol/L (the approximate solubility of CO_2 in a 0.58 M NaCl brine [HUE 10]). The cement contains initially 5.16 mol/L of portlandite (CH) and 3.9 mol/L of Si as jennite (i.e. $C_{1.7}SH_2$) [HUE 10]). The initial porosity is 40%. Concerning the limestone-like reservoir conditions, we set the activity product of calcite, $Q_{C\bar{C}}$, as constant at the interface between the cement paste and the CO_2-saturated brine and equals to the equilibrium constant $K_{C\bar{C}}$, so that at the boundary the pore solution is always saturated with calcite. Concerning the sandstone-like conditions, we set the activity product of calcite as constant and equals to $0.01K_{C\bar{C}}$ to avoid precipitation of calcite at this boundary.

10.4.1. *Sandstone-like conditions*

Duguid *et al.* observed five different regions: an orange ring followed by brown, white, light gray rings and a dark gray core. The orange and brown zones were fully degraded, with little calcium left comparing to the unreacted core. The white ring showed an increase in calcium, corresponding to the presence of the carbonation front. In the light gray ring, portlandite was partially dissolved, reflecting the presence of a dissolution front. The dark gray core was the unreacted zone. The author also notes that the formation of the white and calcium carbonate-rich layer had a protective effect and slowed down the reaction rate.

The simulation results are presented hereafter. The profiles of solid compounds, as well as porosity profiles, are plotted in Figure 10.10. A calcite dissolution occurs due to leaching effect at the boundary. Successive zones, from the boundary to the core, are observed: an amorphous silica gel layer at the boundary, a calcite-rich layer and a dissolution front of C-S-H followed by that of CH. The predicted profiles are quite similar to the simulations of Huet *et al.* [HUE 10] carried out for the same configuration and cementitious system (see Figure 10.12). The calcite precipitation front observed in the experiment [DUG 10] has moved around 1.25 mm into the cement after a 30-day exposure period (vertical dashed lines in Figure 10.10). This is approximatively in agreement with our predictions. We can remark that Huet *et al.* [HUE 10] obtain a prediction of 0.75 mm. This small difference probably comes from the fact that different transport laws are used. Actually, in our model, we consider both diffusion and electric effects for each aqueous species, and we choose a different diffusion coefficient D_i (equation [10.36]) for each species contrary to Huet *et al.* [HUE 10].

A leaching effect is observed here since between the surface and the calcite-rich layer, there is neither calcite nor portlandite or C-S-H, which means that the generated calcite has been dissolved. Thus, at the boundary, the porosity increases to approximately 0.78 due to calcite dissolution.

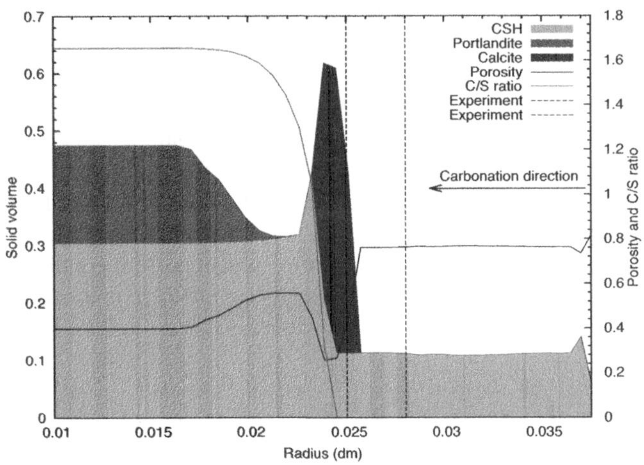

Figure 10.10. *Profiles after 30 days of exposure under sandstone-like conditions (vertical dash lines show the calcite precipitation front assessed by Duguid et al.). For a color version of this figure, see www.iste.co.uk/gpc/geomech.zip.*

10.4.2. *Limestone-like conditions*

Unlike sandstone-conditions, the samples exposed to calcium-saturated brine show no apparent damage in this set of experiments [DUG 10]. Only one light-colored ring with an average depth of 0.58 ± 0.22 mm (vertical dashed lines in Figure 10.11) was observed after a 26-day exposure duration. Duguid does not really explain the meaning of this behavior, for example under sandstone-like condition the samples show rapid acid attack while under limestone-like conditions almost no acid attack takes place. With the help of this numerical model, we can provide a relevant explanation. Our simulations show that we get a full clogging (porosity drops to 0) at the point $r = 0.035$ dm after 10 days of exposure (Figure 10.11) and the calculation then stops due to a zero porosity value. Actually, the generated calcite cannot be dissolved since the brine is already saturated with calcite. This layer protects the inner part of the sample from leaching of calcite and hinders further carbonation. So the carbonation is restricted at the sample surface and the degraded layer depth rapidly stabilizes.

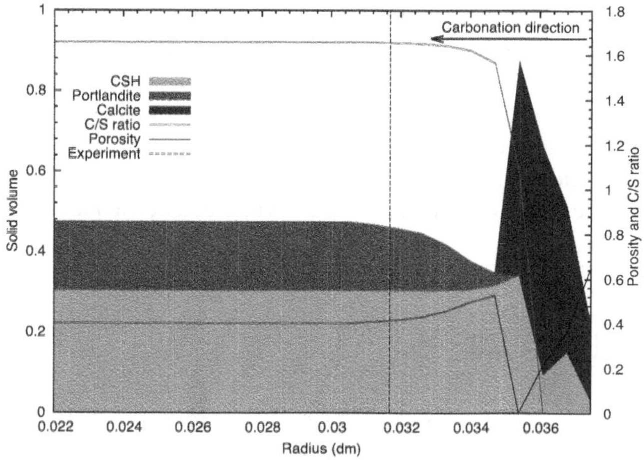

Figure 10.11. *Profiles after 10 days of exposure under limestone-like conditions (vertical dash line shows the carbonation depth assessed by Duguid et al.). For a color version of this figure, see www.iste.co.uk/gpc/geomech.zip.*

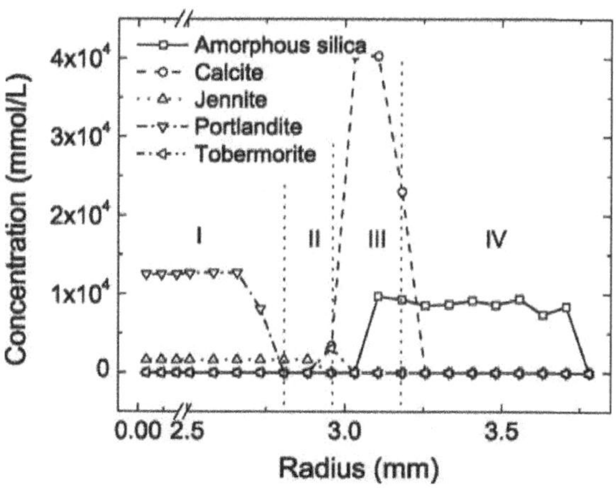

Figure 10.12. *Calculated solid profile after a 30-day exposure time [HUE 10]*

10.4.3. *Study of CO_2 concentration and initial porosity*

In Duguid's experiment, the porosity of the samples is quite high (0.4) and the dissolved CO_2 concentration is rather small (around 0.05 mol/L). Here, we conduct numerical simulations with different initial porosities and CO_2 concentrations to propose a sensitivity analysis of these parameters. The sandstone-like boundary conditions are maintained. A regular 1D axisymmetric configuration with a length of 5 mm divided into 50 nodes is used. The initial CH and C-S-H contents are the same as the previous values for sandstone-like conditions (see Figure 10.10). Figures 10.13–10.15 provide the solid volume and porosity profiles after a 30-day exposure duration with different initial porosities. We can find out that similar successive layers are observed in all conditions, and, as expected, a smaller initial porosity can reduce the developing carbonation ingress kinetics. Figures 10.16–10.17 show the profiles with the same initial porosity and different CO_2 concentrations. A higher CO_2 concentration causes a faster acid attack and a wider calcite-rich layer.

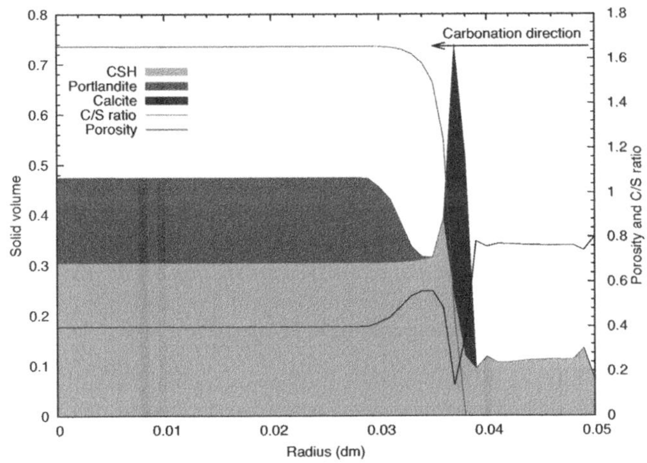

Figure 10.13. *Solid phases and porosity profile after 30 days of exposure. Porosity = 0.4, CO_2 = 0.057 M. For a color version of this figure, see www.iste.co.uk/gpc/geomech.zip.*

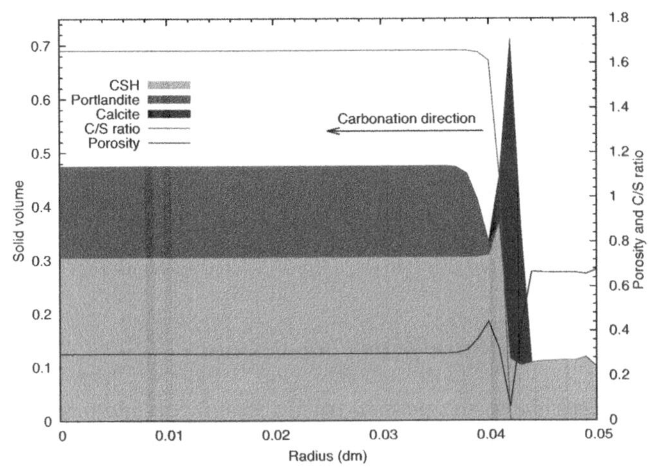

Figure 10.14. *Solid phases and porosity profile after 30 days of exposure. Porosity = 0.3, CO_2 = 0.057 M. For a color version of this figure, see www.iste.co.uk/gpc/geomech.zip.*

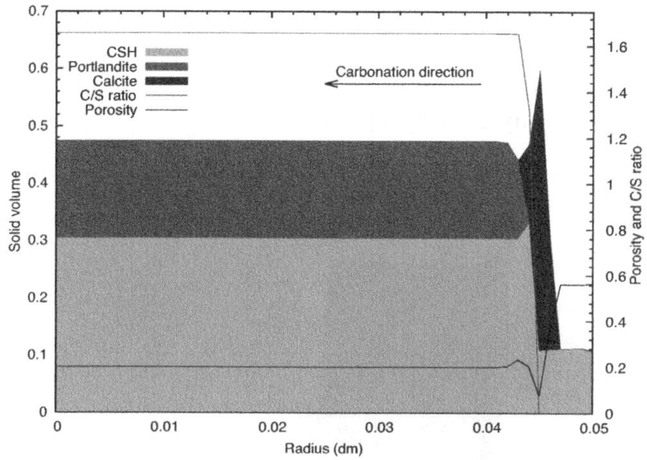

Figure 10.15. *Solid phases and porosity profile after 30 days of exposure. Porosity* $= 0.2$, $CO_2 = 0.057$ *M. For a color version of this figure, see www.iste.co.uk/gpc/geomech.zip.*

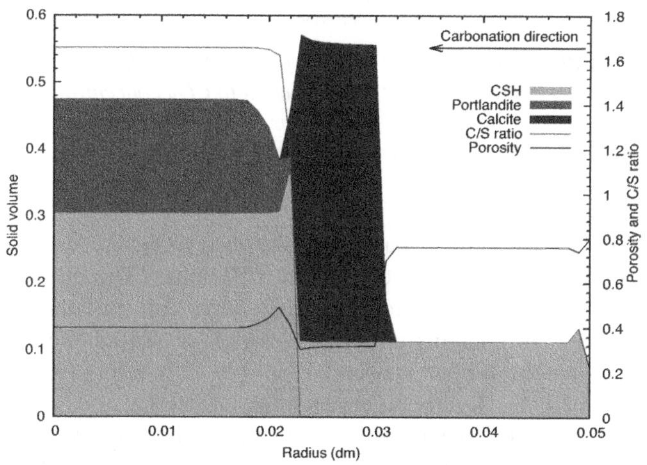

Figure 10.16. *Solid phases and porosity profile after 30 days of exposure. Porosity* $= 0.4$, $CO_2 = 0.35$ *M (equivalent to* CO_2 *concentration at* $T = 293$ *K,* $P = 10$ *bar). For a color version of this figure, see www.iste.co.uk/gpc/geomech.zip.*

10.4.4. *Supercritical boundary conditions*

Rimmelé *et al.* [RIM 08] exposed cement samples to both CO_2-saturated brine and wet scCO_2, under pressure and temperature similar to downhole conditions [RIM 08]. A class G well cement was cured for 72 h at 207 bars and 90°C prior to the experiments. The samples were exposed to wet scCO_2 at 280 bars and 90 °C during varying periods from half a day to 6 months. They estimated the mole fraction of water in the gas phase as 1.8%, which fits fairly well the value predicted by equation [10.2]. They also estimated the dissolved CO_2 in the liquid phase as 1.25 mol/L, which is close to the CO_2 solubility given by equation [10.1].

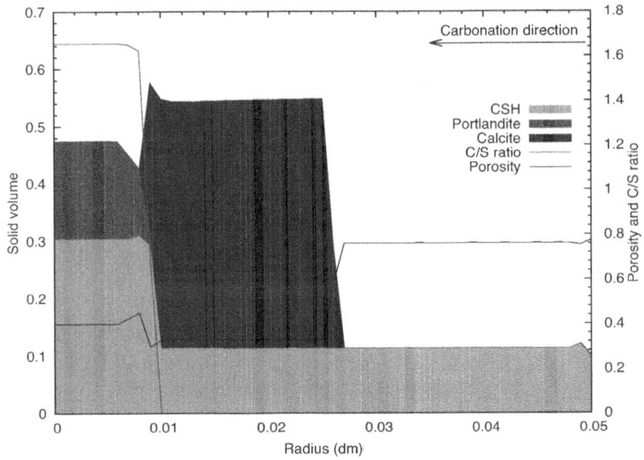

Figure 10.17. *Solid phases and porosity profile after 30 days of exposure. Porosity = 0.4, CO_2 = 0.82 M (equivalent to CO_2 concentration at T = 363 K, P = 100 bar). For a color version of this figure, see www.iste.co.uk/gpc/geomech.zip.*

First we simulate the cement paste exposed to $scCO_2$. We use a cement paste cylinder with 1.25 cm in diameter, divided into 50 nodes. The initial CO_2 pressure equals to 10^{-8} Pa, and increases to 2.8×10^8 Pa in 2 h, just the same as in the experiment. The cement contains initially 5.2 mol/L of portlandite and 3.51 mol/L of Si as jennite. Since the cement was cured for only 72 h before experiment, we use an initial porosity of 47%, slightly higher than that of Duguid and Scherer [DUG 10]. Since the sample is not in contact with water, we set the activity product of calcite ($Q_{C\bar{C}}$) at the boundary of cement paste equal to the equilibrium constant $K_{C\bar{C}}$, so that at the boundary there is no leaching of calcium due to diffusion effect. The liquid pressure is set to 10^7 Pa.

Rimmelé has observed a pH drop from an initial value of 13 to 6–7 after CO_2 exposure, while we predict a drop from 12.4 to 5 as shown in Figure 10.18. The $scCO_2$ has not entered the cement sample during time, since the sample is initially saturated and the liquid pressure increases along with gas pressure. From the rim to the center, a carbonated zone, a carbonation front (rich in calcite), a dissolution front and the inner part were predicted as the experiment observation; see Figures 10.20 and 10.21. The measured alteration front increases with time, 1–2 mm after 2 days and 5–6 mm after 3 weeks of attack. After 6 weeks, the front has reached the central part of the cement sample. From the porosity profiles (Figure 10.20), we can see that the calculated alteration front, where the porosity changes dramatically, reaches to 1–2 mm after 2 days and 4–5 mm after 3 weeks, and the front reaches the center after 6 weeks in our simulation.

Despite the good corresponding between the developing speed of the alteration front, we do not predict much dissolution of calcite at the surface of the sample. This is probably due to our boundary condition, setting $Q_{C\bar{C}}/K_{C\bar{C}} = 1$ at the surface preventing from calcite dissolution. But in the experiment, there is still some calcite dissolution observed, although limited comparing with the samples in CO_2-saturated water.

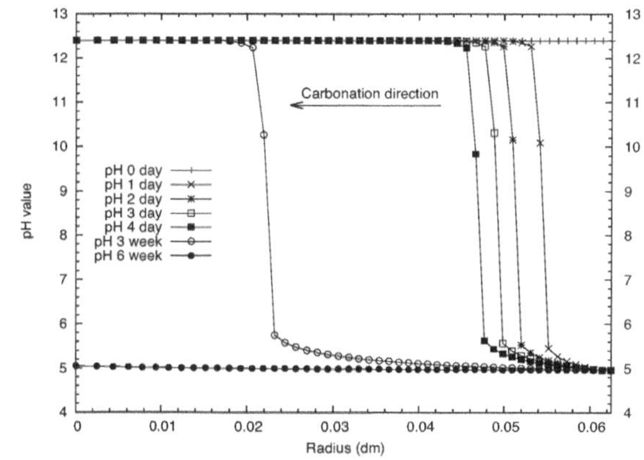

Figure 10.18. *The calculated* pH *profile from 1 day to 6 weeks*

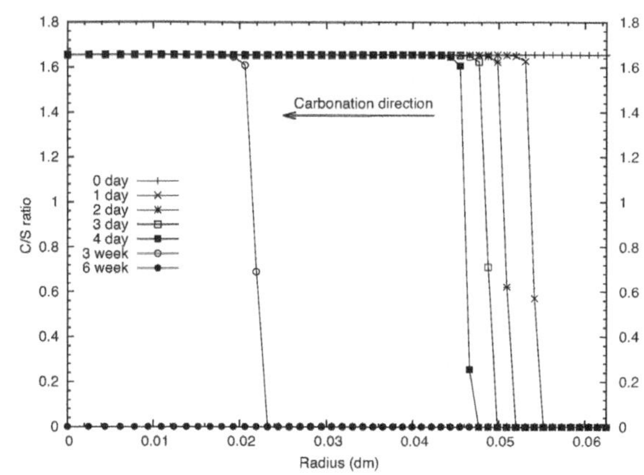

Figure 10.19. *The calculated C/S ratio of CSH from 1 day to 6 weeks*

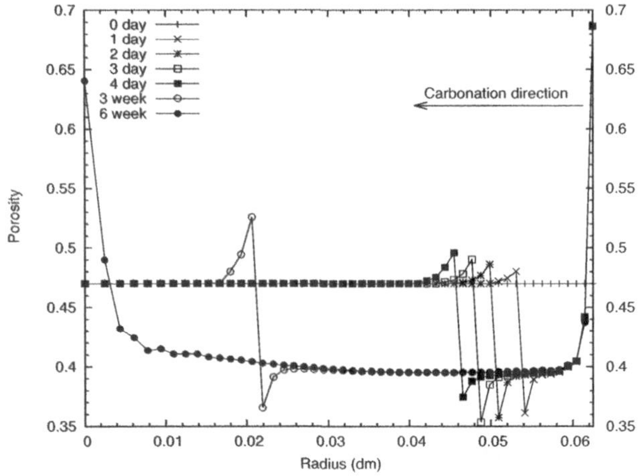

Figure 10.20. *The calculated porosity profile from 1 day to 6 weeks*

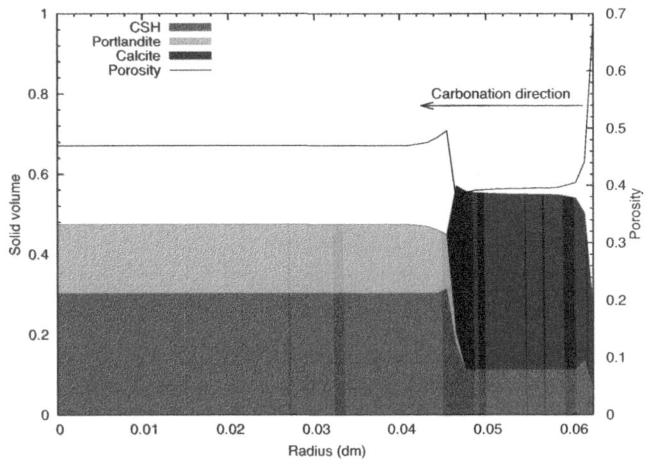

Figure 10.21. *The calculated solid profile at 4 days*

10.5. Conclusion

To predict CO_2 penetration through oil well cement during long-term storage, a reactive transport model is built. This model is able to simulate the cement paste behavior in contact with both water and supercritical CO_2 ($scCO_2$). The considered solid and aqueous species, the transport equations and the simultaneous method used to couple the transport equations and the chemical reactions are introduced. The simulations are compared with some experimental observations. The various zones observed in the experiment, for example a porous silica gel layer at the boundary, a calcite-rich layer, the dissolution front where C-S-H starts to dissolve followed by the carbonation front where CH is partially dissolved, are simulated. By using a transport law suitable for cement-based materials, as well as considering the diffusion and the electric effects for each aqueous species, this model can provide quite a good prediction of the carbonation kinetics of a cement paste exposed to a CO_2-saturated brine for sandstone-like and limestone-like conditions, as well as to $scCO_2$. Some experimentally observed phenomena can also be explained by the model in terms of calcite front, dissolution front, porosity profile, etc. Further improvement of the model should include the humidity of $scCO_2$ and the effect of high temperatures.

10.6. Acknowledgment

The authors would like to thank the General Council of "Seine et Marne" for their financial support. The authors also express their gratitude to Bruno Capra and Olivier Poupard, from Oxand (http://www.oxand.com/), for their advice and help.

10.7. Bibliography

[BAR 07] BAROGHEL-BOUNY V., "Water vapour sorption experiments on hardened cementitious materials: Part I: essential tool for analysis of hygral behaviour and its relation to pore structure", *Cement and Concrete Research*, vol. 37, no. 3, pp. 414–437, 2007.

[CAR 07] CAREY J.W., WIGAND M., CHIPERA S.J., WOLDEGABRIEL G., PAWAR R., LICHTNER P.C., WEHNER S.C., RAINES M.A., GUTHRIE G.D., *et al.*, "Analysis and performance of oil well cement with 30 years of CO_2 exposure from the Sacroc unit, West Texas, USA", *International Journal of Greenhouse Gas Control*, vol. 1, no. 1, pp. 75–85, 2007.

[DUG 09] DUGUID A., "An estimate of the time to degrade the cement sheath in a well exposed to carbonated brine", *Energy Procedia*, vol. 1, no. 1, pp. 3181–3188, 2009.

[DUG 10] DUGUID A., SCHERER G.W., "Degradation of oilwell cement due to exposure to carbonated brine", *International Journal of Greenhouse Gas Control*, vol. 4, no. 3, pp. 546–560, 2010.

[FEN 98] FENGHOUR A., WAKEHAM W.A., VESOVIC V., "The viscosity of carbon dioxide", *Journal of Physical and Chemical Reference Data*, vol. 27, no. 1, pp. 31–44, 1998.

[GRE 65] GREENBERG S.A., CHANG T.N., "Investigation of the colloidal hydrated calcium silicates. ii. solubility relationships in the calcium oxidesilicawater system at 25°", *The Journal of Physical Chemistry*, vol. 69, pp. 181–188, 1965.

[HUE 10] HUET B.M., PREVOST J.H., SCHERER G.W., "Quantitative reactive transport modeling of portland cement in CO_2-saturated water", *International Journal of Greenhouse Gas Control*, vol. 4, no. 3, pp. 561–574, 2010.

[JEN 04] JENNINGS H.M., "Colloid model of c-s-h and implications to the problem of creep and shrinkage", *Materials and Structures*, vol. 37, pp. 59–70, 2004.

[KUL 01] KULIK D.A., KERSTEN M., "Aqueous solubility diagrams for cementitious waste stabilization systems: Ii, end-member stoichiometries of ideal calcium silicate hydrate solid solutions", *Journal of the American Ceramic Society*, vol. 84, no. 12, pp. 3017–3026, 2001.

[LOT 08] LOTHENBACH B., MATSCHEI T., MÖSCHNER G., GLASSER F.P., "Thermodynamic modelling of the effect of temperature on the hydration and porosity of portland cement", *Cement and Concrete Research*, vol. 38, no. 1, pp. 1–18, 2008.

[MUA 76] MUALEM Y., "Hysteretical models for prediction of the hydraulic conductivity of unsaturated porous media", *Water Resources Research*, vol. 12, no. 6, pp. 1248–1254, 1976.

[REA 90] REARDON E.J., "An ion interaction model for the determination of chemical equilibria in cement/water systems", *Cement and Concrete Research*, vol. 20, pp. 175–192, 1990.

[REA 92] REARDON E.J., "Problems and approaches to the prediction of the chemical composition in cement/water systems", *Waste Management*, vol. 12, pp. 221–239, 1992.

[RED 49] REDLICH O., KWONG J.N.S., "On the thermodynamics of solutions. v. an equation of state. fugacities of gaseous solutions", *Chemical Reviews*, vol. 44, pp. 233–244, 1949.

[RIM 08] RIMMELÉ G., BARLET-GOUÉDARD V., PORCHERIE O., GOFFÉ B., BRUNET F., "Heterogeneous porosity distribution in portland cement exposed to CO_2-rich fluids", *Cement and Concrete Research*, vol. 38, no. 8–9, pp. 1038–1048, 2008.

[SPY 03] SPYCHER N., PRUESS K., ENNIS-KING J,. "CO_2-H_2O mixtures in the geological sequestration of CO_2. i. assessment and calculation of mutual solubilities from 12 to 100 c and up to 600 bar", *Geochimica and Cosmochimica Acta*, vol. 67, no. 16, pp. 3015–3031, 2003.

[THI 06] THIERY M., BAROGHEL-BOUNY V., DANGLA P., VILLAIN G., "Numerical modeling of concrete carbonation based on durability indicators", *ACI Materials Journal*, vol. 234, pp. 765–780, 2006.

[THI 12] THIERY M., BAROGHEL-BOUNY V., MORANDEAU A., DANGLA P., "Impact of carbonation on the microstructure and transfer properties of cement-based materials", *Transfert 2012*, Ecole Centrale de Lille, pp. 1–10, 2012.

[THO 03] THOENEN T., KULIK D., Nagra/psi chemical thermodynamic data base 01/01 for the gem-selektor (v. 2-psi) geochemical modeling code: Release 28-02-03, Technical report, PSI Technical Report TM-44-03-04 about the GEMS version of Nagra/PSI chemical thermodynamic database 01/01, 2003.

[TOG 98] TOGNAZZI C., Couplage fissuration-dégradation chimique dans les matériaux cimentaires : caractérisation et modélisation, PhD thesis, INSA Toulouse, 1998.

[VAN 80] VAN GENUCHTEN M.T., "A closed-form equation for predicting the hydraulic conductivity of unsaturated soils", *Soil Science Society of America Journal*, vol. 44, no. 5, pp. 892–898, 1980.

Chapter 11

Chemo-Poromechanical Study of Wellbore Cement Integrity

11.1. Introduction

The safety assessment of carbon dioxide storage sites requires refined analyses accounting for the multi physical phenomena occurring in the different geological and man-made materials that can be encountered in the storage complex. Even though reactive transport phenomena play a significant role in these applications, the geomechanics of CO_2 storage has to be accounted for. However, this is not the case in the majority of studies available to date. One of the main reasons is that the difficulties arise when looking at the interaction between chemical and poromechanical phenomena. Because of the complexity of this highly coupled problem, and before getting into difficult numerical analyses, we have to set sound foundations to build a rigorous theoretical framework able to model highly coupled and nonlinear phenomena involved when analyzing CO_2 geological storage. Poromechanics is seen as a facilitating tool that permits us to encompass the different couplings and to provide this sound framework.

When dealing with the numerical resolution of such coupled problems, one faces the lack of fully coupled codes involving both reactive transport and poromechanical (or geomechanical) problems. One of the simplest solutions is to abruptly couple existing codes but in this case, the resolution of the problem cannot be monolithic and the problem has to be decoupled, at least partly. This solution presents the clear advantage of limiting the required numerical developments to their strict minimum, to benefit from the richness of each code and to minimize the cost of calculations.

Chapter written by Jean-Michel PEREIRA and Valérie VALLIN.

However, in some cases, when the physical couplings are strong or when high gradients are expected for instance, such resolutions using weak couplings can lead to incorrect results. On the contrary, a fully coupled resolution ensures that no numerical artifact is introduced in the solution, which cannot be guaranteed *a priori* by a sequential resolution alternating the resolutions of the reactive transport problem and of the mechanical problem. Commercial codes enabling such a fully coupled resolution strategy are almost non-existent to date. It must be recognized that the calculation costs (in terms of computation time and required hardware), if such tools exist, could be very high. One solution in a first stage is to identify the main chemical phenomena playing a significant role in the overall problem to solve and not to neglect such significant phenomena. This simplified version of the studied problem would then be implemented and solved monolithically. A further interest of such a methodology is that it furnishes reference solutions that could then be used to benchmark other codes used to solve more sophisticated problems.

In this chapter, focus is placed on the behavior of the cement sheath constituting the injection well. This cement will indeed be subjected to rapid reactions provoked by the presence of carbon dioxide in the storage complex. Such a focus is legitimized by the fact that wells are seen as one of the main discontinuities in the storage sealing complex that could turn into preferential pathways for CO_2 leaks toward the biosphere. Even though CO_2 resistant cements will be used, abandoned wells could be present in the storage site and be in contact with CO_2. A chemo-poromechanical framework for chemo-poromechanical modeling of dissolution/precipitation phenomena is thus presented. Its application to cement carbonation and its implementation into a numerical code is then discussed. A simplified analysis, representative of CO_2 geological storage, is finally performed to assess the capabilities of the numerical tool thus developed.

11.2. Poromechanics of cement carbonation in the context of CO_2 storage

11.2.1. *Context and definitions*

We consider a porous solid whose matrix may react with the fluid(s) contained in its porous network. The reactions considered here include dissolution of solid minerals constituting the matrix and precipitation of solid constituents within the available porous space. As discussed later, this topic presents formal similarities with the case of partially saturated porous solids, and some of the concepts introduced in recent studies of this latter case are recalled hereafter.

First of all, let us define the thermodynamic system under consideration. A partially saturated porous medium comprises a solid skeleton in contact, through interfaces having their own energy, with two fluid phases, typically a gas phase and a liquid phase[1]. The thermodynamics of this solid skeleton can be addressed by considering successively three systems. The first system is the porous medium itself, as just depicted. This is an open system exchanging gas and liquid mass with the surroundings. The second system is obtained by removing the bulk gas and liquid phases from the medium. This is done for convenience and with no loss of information because the thermodynamics of the bulk fluids is known separately and because the focus is laid on the behavior of the porous solid[2]. Since this second system does not include any more of the fluid phases, it is formed of the solid phase and the interfaces. The removal of the bulk fluid phases does not mean that the system is no longer subjected to the pore pressures. This system is actually a closed system that is loaded by the total external stress and the pore pressures still exerting through the interfaces on the solid. In the following, this system is called the apparent solid skeleton, since the interfaces have their own energy. As a result they also have to be removed to define the actual solid skeleton, the constitutive equations of which we are looking for. We call it the solid skeleton in the following. In short, three systems are considered: the porous medium (solid skeleton + interfaces + bulk fluid phases), the apparent solid skeleton (solid skeleton + interfaces), and the solid skeleton.

In the reference configuration, an infinitesimal representative elementary volume (REV) of the porous medium has a volume $d\Omega_0$. In the current configuration, this volume is denoted as $d\Omega_t$. The pore volume contained in the REV in the current configuration is given by $\phi\, d\Omega_0$, where ϕ is the Lagrangian porosity of the medium. As seen in Figure 11.1, the solid matrix is taken as the solid fraction corresponding to a reference state where no chemical alteration occurred. Any solid precipitation is then assumed to occur in the porous space. The overall porosity ϕ can then be split as follows:

$$\phi = \phi_F + \phi_P \qquad [11.1]$$

where $\phi_F\, d\Omega_0$ is the volume occupied by the pore fluid in the current configuration and $\phi_P\, d\Omega_0$ corresponds to the volume of the precipitated minerals (or any non-wetting phase in a more general case) in the current configuration. ϕ_F and ϕ_P are called partial porosities and correspond to the porous space occupied by the two species considered

1 Actually, in a more general case, the porous space is filled by wetting fluid, which could be water, and a non-wetting phase, which could be oil or crystals such as water ice (see for instance [COU 05]) or precipitated minerals (such as calcium carbonates in the present study). In the case of solid phases (ice, crystals), the authors reckon that things can be more complicated if the energetic interactions between the filling solid phase and the solid skeleton involve not only isotropic stresses but also shear stresses. This case was avoided in [COU 05] by invoking the presence of a thin film of super-cold liquid water.

2 Removing the fluids is possible in classical poromechanics because it is assumed that they do not physically interact with the solid. When this assumption is not relevant, a new poromechanical framework has to be used. Interested readers can refer to [VAN 10] where the authors account for the effects of adsorption of CO_2 on the swelling of a coal matrix.

here. Equation [11.1] introduces a split in the porous volumes occupied by two distinct phases. In this sense and as stated previously, a relatively direct correspondence with poromechanics of partially saturated media holds.

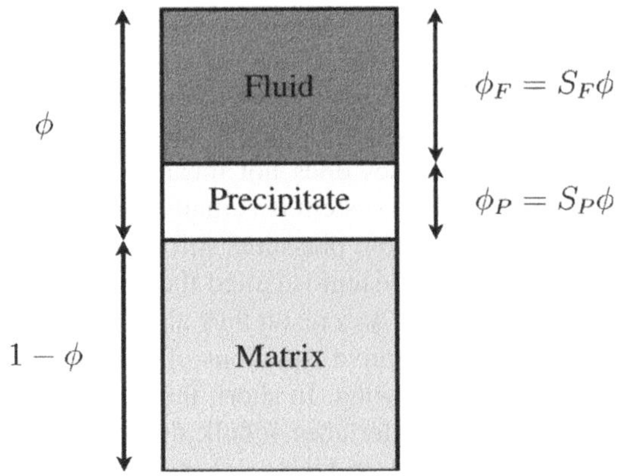

Figure 11.1. *Volume fractions*

Following [COU 04] and [COU 10], and besides the use of Lagrangian porosities, we introduce Lagrangian degrees of saturation attached to the phases filling the porous space. Working with these Lagrangian quantities instead of their Eulerian counterparts present several advantages that make life easier when dealing with thermodynamics of porous media. These advantages are now briefly exposed.

When considering reversible processes, the balance of free energy of the apparent solid skeleton[3] is written as follows:

$$\sigma_{ij}d\epsilon_{ij} + p_{nW}d\phi_{nW} + p_W d\phi_W - d\psi = 0 \qquad [11.2]$$

The indices nW and W stand for the non-wetting and wetting phase, respectively. Equation [11.2] corresponds to the work input derived by [HOU 97]. It is perfectly correct but unable to provide a strain energy balance separately from the surface energy balance. Indeed, both deformation processes and fluid invasion processes will induce variations of the partial porosities $\phi_W = s_W\phi$ and $\phi_{nW} = s_{nW}\phi$, but the energy balance expressed by equation [11.2] does not permit us to distinguish the corresponding contributions to the change of free energy of the solid skeleton.

3 The apparent solid skeleton is assumed to be chemically inert here, and no physical interaction between the saturating fluids and the solid phase is considered. Doing so, the present formulation is equivalent to the derivations obtained for partially saturated porous media.

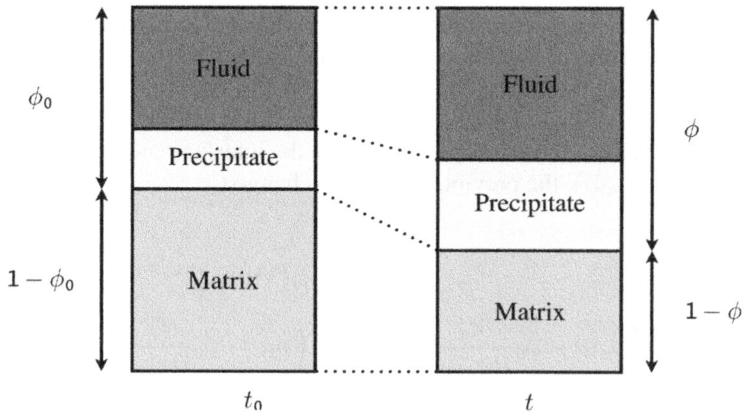

Figure 11.2. *Evolution of volume fractions during dissolution/ precipitation processes*

Let us now introduce the Lagrangian degrees of saturation S_{nW} and S_W defined by:

$$\phi_\alpha = S_\alpha \phi_0 + \varphi_\alpha; \quad \alpha = \{W, nW\} \qquad [11.3]$$

where φ_α stands for the change due to deformation of the partial porosity associated with the phase α. Splitting the free energy of the skeleton as follows:

$$\psi = \psi_s + \phi_0 U \qquad [11.4]$$

where ψ_s is the free energy of the solid skeleton and U is the free energy of the interfaces, and using equation [11.3] enables us to write the energy balance as follows:

$$p_{nW} - p_W = -\frac{dU}{dS_W} \qquad [11.5]$$

$$\sigma_{ij} d\epsilon_{ij} + p_{nW} d\varphi_{nW} + p_W d\varphi_W - d\psi_s = 0 \qquad [11.6]$$

It is assumed here that the energy of the interfaces U depends only on the degrees of saturation and not on deformation, that is $U = U(S_W)$[4]. Equation [11.5] now corresponds to the energy balance of the interfaces and this relation corresponds to the water retention curve of the porous solid. Equation [11.6] corresponds to the energy balance of the solid skeleton. The comparison between equations [11.2] and [11.6] is interesting in this sense. In the latter, the work input due to the fluid pressures is associated with the component of porosities related to deformational processes only

4 It is clear that a more general form would be $U = U(S_W, S_{nW})$ but it can be simplified because both degrees of saturation are linked through the relation $S_{nW} + S_W = 1$.

(φ_α). If we consider an invasion process (fluid drainage or imbibition) into a non-deformable solid skeleton, equation [11.6] vanishes and all the energy changes are captured by equation [11.5].

If we want to release the assumption made on the interface energy and assume that $U = U(S_W, \varphi_W, \varphi_{nW})$, the previous equations change to:

$$p_{nW} - p_W = -\frac{\partial U}{\partial S_W} \qquad [11.7]$$

$$\sigma_{ij} d\epsilon_{ij} + \left(p_{nW} - \phi_0 \frac{\partial U}{\partial \varphi_{nW}}\right) d\varphi_{nW} + \left(p_W - \phi_0 \frac{\partial U}{\partial \varphi_W}\right) d\varphi_W - d\psi_s = 0 \qquad [11.8]$$

If we now compare equations [11.6] and [11.8], the role played by the deformation of the interfaces on the work input associated with the fluid pressures appears. This corrected or apparent pressure can be expressed as $p_\alpha - \phi_0 \frac{\partial U}{\partial \varphi_\alpha}$. It is worth noting that this effect was accounted for in [COU 02, DAN 02] and used by [PER 05] in the integral term of the equivalent pore pressure π:

$$\pi(s_W, \phi) = s_{nW} p_{nW} + s_W p_W - \frac{2}{3} \int_{s_W}^{1} p_c(s_W, \phi) ds_W \qquad [11.9]$$

The theoretical framework previously described is now extended to account for chemical reactions (dissolution/precipitation phenomena). Since we are interested in the chemo-poromechanical couplings taking place within the cement sheath in the context of CO_2 geological storage, it is necessary to inspect the main chemical reactions that could occur and how these reactions will affect the poromechanical behavior of the solid skeleton. Once done, these reactions will have to be introduced in a consistent theoretical framework. Both objectives are the subject of the following two sections.

11.2.2. *Chemical reactions*

The injection of carbon dioxide into the reservoir will acidify not only the water present in the vicinity of the CO_2 plume. In this study, the focus is laid on the region of the triple point, separating the injection well, the reservoir rock (deep saline aquifer, for instance) and the caprock (low-porosity and low-permeability rock such as argilites, for instance). The CO_2 plume will reach this zone relatively rapidly and even if the pressure of the gaseous CO_2 will remain below the entry pressure of the caprock, it is expected that dissolved CO_2 will diffuse within the caprock. The presence of the CO_2 induces an acidification of the pore water and will imply the carbonation of the geomaterials and cement. With respect to the cement, this carbonation consists of the lixiviation or alteration of the mineral phases constituting the cement. Some reaction products (such as calcium carbonates) will precipitate in the available porosity. As result, these reactions will have an impact on the porosity of the material and on its mechanical (stiffness and strength) and transport (permeability and retention) properties.

To simplify the picture, we will consider only two hydrates in the cement paste: portlandite, $Ca(OH)_2$, noted as CH in cement industry and calcium silicate hydrates, noted as C–S–H. According to [THI 05], predominant mechanisms of cement carbonation are related to these two hydrates. The chemical reactions accounted for in the present study read as follows:

$$\begin{array}{lrcl} R_1 & Ca(OH)_2 + CO_2 & \longrightarrow & CaCO_3\downarrow + H_2O \\ R_2 & 2.5\,C{-}S{-}H_{1.6} + CO_2 & \longrightarrow & 2.5\,C{-}S{-}H_{1.2} + CaCO_3\downarrow + 1.3\,H_2O \\ R_3 & 2.5\,C{-}S{-}H_{1.2} + CO_2 & \longrightarrow & 2.5\,C{-}S{-}H_{0.8} + CaCO_3\downarrow + 1.3\,H_2O \\ R_4 & 1.25\,C{-}S{-}H_{0.8} + CO_2 & \longrightarrow & 1.25\,SiO_{2(am)} + CaCO_3\downarrow + 1.925\,H_2O \end{array}$$

The symbol $\downarrow$ indicates the precipitation of the product. It can be seen that these reactions produce calcium carbonates (mainly calcite)[5] and amorphous silica. Note that the carbonation of CH is relatively simple (reaction R_1). On the contrary, C–S–H carbonation leads to a progressive loss of density linked to a decalcification of their structure and ends with the formation of amorphous silica SiO_2 (see reactions R_2 to R_4)[6]. The subscripts 1.6, 1.2 and 0.8 correspond to the C/S density ratio in the C–S–H phase.

It is worth noting that reaction kinetics are neglected in the following. This assumption is based on previous works on dissolution/precipitation processes in cements (see [MAI 99] for instance) and on simulations performed with the reactive transport code TOUGHREACT [XU 06], which showed that the characteristic times of the reactions considered in this study are much lower than the characteristic time of CO_2 diffusion in the liquid phase, which will be the limiting process in the studied problem[7]. Apart from being instantaneous, we will further assume that the reactions

5 Among the different crystallographic forms of calcium carbonates, calcite is seen here as the most stable and thus the most probable crystal produced by the carbonation reactions considered here. This assumption permits us to simplify the model formulation, but the authors recognize that other polymorphs such as aragonite could also be formed.

6 It should be noted that the silica produced is likely to form under an hydrated form but this is not accounted for here. This does not restrict the generality of the proposed methodology.

7 It is worth noting that the problem studied here will not involve strong pressure gradients so that advective transport of CO_2 will not play a key role.

R_i are complete and successive. We can write these four reactions in the following synthetic form:

$$a_{1,R_i} M^S_{1,R_i} + \mathrm{CO_2} \longrightarrow a_{2,R_i} M^S_{2,R_i} + a_{\mathrm{H_2O},R_i} \mathrm{H_2O} + \mathrm{CaCO_3} \downarrow; \quad i \in \{1,4\} \qquad [11.10]$$

where a_{j,R_i} is the stoechiometric coefficient of the mineral M^S_{j,R_i} in the reaction R_i, with $j = 1$ for a reactant and $j = 2$ for a product in the same reaction.

If we consider a continuous supply of CO_2-rich water, it is clear that the carbonates produced from reactions (R_1) to (R_4) will eventually dissolve. However, in the context of CO_2 storage, and considering the typical target geological formations for the reservoir and its cap, it is likely that the *in situ* pore water will contain relatively important amounts of dissolved calcium, which will act as a buffer, thus impeding a complete dissolution of the produced calcite. Here, it is implicitly assumed that the *in situ* pore water is saturated by calcium ions in the initial state and that calcium carbonates will precipitate using the calcium leached from the cement matrix.

The in-pore fluid solution is a mixture containing several species: the water being the solvent and several solutes coming from the dissolved matrix, dissolved carbonates, and dissolved CO_2. Let $n_i \times d\Omega_0$ be the number of moles of species i present in a unit volume of the porous medium $d\Omega_0$, so that n_i is the apparent molar density of species i. The isothermal Gibbs–Duhem equality assuming chemical equilibrium can be written relative to the in-pore solution and to the carbonate crystal so that:

$$-\phi_C \frac{dp_C}{dt} + n^S_C \frac{d\mu^S_C}{dt} = 0 \qquad [11.11]$$

$$-\phi_F \frac{dp_F}{dt} + \sum_{i=\alpha,w} n^F_i \frac{d\mu^F_i}{dt} = 0 \qquad [11.12]$$

where p_C and p_F are the pressures of the calcite crystal and the in-pore fluid, respectively, and μ^S_C and μ^F_α stand for the molar chemical potentials of calcite and aqueous species α, respectively. In equation [11.12], the energy transmitted through shear stresses between the carbonate crystal and the surrounding solid matrix is neglected with respect to that transmitted through the normal stress.

Chemical equilibrium conditions between the fluid, crystals, and cement matrix are now briefly described. Since the reactions of precipitation of calcium carbonate and dissolution of cement hydrates are assumed to be instantaneous, the thermodynamical equilibrium between dissolved species in the fluid and the corresponding solid mineral is valid at any time and in every REV. In the case of calcium carbonate, this amounts to saying that the chemical potential of carbonate crystals is in equilibrium with the chemical potential of dissolved crystals:

$$\mu_C = \mu_{\mathrm{Ca}^{2+}} + \mu_{\mathrm{CO}_3^{2-}} \qquad [11.13]$$

Finally, changes in the molar quantity of any species, denoted as dn_i/dt, is due to both exchanges of matter with surrounding elementary volumes and to chemical reactions. Assuming that molar transport of solid minerals is not significant, conservation equations can be expressed such as:

$$\frac{dn^S_{M_i,C}}{dt} = \mathring{n}^S_{M_i,C} \qquad [11.14]$$

$$\frac{dn^F_{\alpha,w}}{dt} = -\mathrm{div}\boldsymbol{w}^F_{\alpha,w} + \mathring{n}^F_{\alpha,w} \qquad [11.15]$$

with $\mathring{n}^S_{M_i,C}$ and $\mathring{n}^F_{\alpha,w}$ being the molar variations of solid and dissolved species due to the chemical reactions and $\boldsymbol{w}^F_{\alpha,w}$ the molar transport of species α in the fluid phase. This latter will be made explicit in the following section (see equation [11.21]).

11.2.3. *Chemo-poromechanical behavior*

In the previous section, the chemical background has been depicted. We will now focus on the poromechanical behavior of the cement and more particularly on the treatment of the chemical effects on the mechanical behavior. As discussed earlier, a distinction is made between the porosity filled by the fluid solution (water and dissolved species) and that filled by the precipitated crystals. In that way, the present study borrows several tools developed for the poromechanics of partially saturated media[8]. This distinction is relevant to precisely follow the processes of dissolution/precipitations of the cement matrix and the calcite. These partial porosities are linked through the following relation:

$$\phi = \phi_F + \phi_C \qquad [11.16]$$

where ϕ_F is the porosity filled by the fluid solution (which corresponds to the "real" connected porous network) and ϕ_C corresponds to the fraction of the total volume occupied by the precipitated calcium carbonates. With such a definition, $(1 - \phi)d\Omega_0$ corresponds to the volume of the cement skeleton (excluding the crystals) in the actual

8 This is the case for the use of two partial porosities, each one filled by a phase at a given pressure. This is also the case for the mass transport of the fluid phase here, which takes place in the partial porosity associated with the fluid mixture as the advective transport of the wetting phase occurs in the porosity it fills in partially saturated media.

configuration. This solid matrix may be composed of inert minerals, amorphous silica, and portlandite and C–S–H that still have not reacted with CO_2.

Several porosity changes may be identified, some being the result of chemical processes, the others from processes of deformation of the porous medium. Let ϕ_L represent porosity changes due to the leaching of cement matrix and ϕ_P the porosity change due to carbonates precipitation. Let φ_F and φ_C be the deformation of the porosity filled by the fluid phase and by the calcite phase, respectively. The partial porosities involved in equation [11.16] can thus be rewritten as follows:

$$\phi_F = \phi_0 + \phi_L - \phi_P + \varphi_F \qquad [11.17]$$

$$\phi_C = \phi_P + \varphi_C \qquad [11.18]$$

Here, it is assumed that all the calcium carbonate possibly present in the cement in its initial state pertains to its solid skeleton so that the initial porosity ϕ_0 is a "real" porosity. ϕ_C is introduced as a measure of the changes in carbonates content with respect to this initial state.

Let $\boldsymbol{\epsilon}$ be the overall infinitesimal strain tensor of $d\Omega_0$ and $\boldsymbol{\sigma}$ the overall stress tensor to which the considered system is subjected. Combining the first and second laws of thermodynamics, the isothermal Clausius–Duhem inequality for the porous medium composed of the solid skeleton and the phases filling its porosity including their interfaces is obtained (see [COU 04]) by:

$$\sigma_{ij} d\epsilon_{ij} - \sum_{i=\alpha,w} \mu_i \nabla \cdot \boldsymbol{w}_i - d\psi \geq 0 \qquad [11.19]$$

where ψ is the free energy of the system that can be split into the free energy of the solid skeleton ψ_s (see previous comments), of the precipitates $\psi_C = n_C \mu_C - \phi_C p_C$, and of the fluid constituents $\psi_F = \sum_{k=\alpha,w} n_k \mu_k - \phi_F p_F$. Assuming that the solid skeleton, the precipitated minerals and the fluid solution are in equilibrium at all times, it can be written that:

$$\sigma_{ij} d\epsilon_{ij} - \sum_{M_i} \mu_{M_i} \delta n_{M_i} + \sum_{k=F,C} p_k d\phi_k - d\psi_s \geq 0 \qquad [11.20]$$

which now concerns the apparent solid skeleton only.

In equation [11.19], $\boldsymbol{w}_i^F$ is the vector of molar transport of the aqueous species i. Thus, $-\mathrm{div}\boldsymbol{w}_i^F$ is the rate of moles of aqueous species i externally supplied to the infinitesimal porous element $d\Omega_0$ by its contiguous elements. Considering Fick's law and Darcy's law for the diffusive and advective transport of species i, $\boldsymbol{w}_i^F$ can be expressed as:

$$\boldsymbol{w}_i^F = -d_{\text{eff}}\, \mathbf{grad} C_i^F - C_i^F \frac{\kappa}{\eta_{\text{vis}}} \mathbf{grad} p_F \qquad [11.21]$$

with d_{eff} being the effective diffusion coefficient, κ the intrinsic permeability, η_{vis} the dynamic viscosity of the fluid phase, C_i^F the molar concentration of the species i and p_F the fluid pressure.

From Clausius–Duhem inequality [11.19] and assuming isothermal conditions, the constitutive equations of isotropic linear poroelasticity can be deduced (see [COU 04] for more details):

$$\boldsymbol{\sigma} - \boldsymbol{\sigma}_0 = \left(K - \frac{2}{3}G\right)(\varepsilon - \varepsilon_0)\mathbf{1} + 2G\,(\boldsymbol{\epsilon} - \boldsymbol{\epsilon}_0) - \sum_{K=F,C} b_K(p_K - p_{K,0})\mathbf{1} \quad [11.22]$$

$$\varphi_J - \varphi_{J,0} = b_J\,(\varepsilon - \varepsilon_0) + \sum_{K=F,C} \frac{p_K - p_{K,0}}{N_{JK}};\ J = F, C \quad [11.23]$$

where $\varepsilon = \text{tr}(\boldsymbol{\epsilon})$ is the volumetric deformation, K and G are the bulk modulus and the shear modulus of the empty porous solid, respectively, and b_J and N_{JK} are the generalized Biot coefficient and the generalized poroelastic coupling moduli, respectively.

Since the chemical reactions induce porosity changes[9], the mechanical moduli of the porous medium will be affected. To account for these effects, the three-phase self-consistent micromechanical model presented in [FEN 98] under the assumption of local isotropy is used. The effective bulk modulus of the porous medium can thus be written as:

$$K = \frac{4G_m K_m (1 - \phi)}{4G_m + 3K_m \phi} \quad [11.24]$$

where K_m and G_m are the bulk and shear modulii of the solid matrix, respectively. The effective shear modulus G can be obtained, thanks to G_m, from the following quadratic equation of which it is the positive root:

$$\lambda_2 \left(\frac{G}{G_m}\right)^2 + \lambda_1 \left(\frac{G}{G_m}\right) + \lambda_0 = 0 \quad [11.25]$$

with λ_0, λ_1 and λ_2 being the coefficients linked to the effective porosity ϕ and Poisson's ratio ν of the porous medium:

9 Mechanical and hydraulic loadings will also lead to porosity changes but in the targeted application, these changes are expected to be far less important than changes due to chemical reactions.

$$
\begin{aligned}
\lambda_2 &= 2(4-5\nu)(7+5\nu)\phi^{\frac{10}{3}} - 25(\nu^2-7)\phi^{\frac{7}{3}} - 252\phi^{\frac{4}{3}} \\
&\quad +50(7-12\nu+8\nu^2)\phi + 4(7-10\nu)(7-5\nu) \\
\lambda_1 &= -(7+5\nu)(1-5\nu)\phi^{\frac{10}{3}} + 50(\nu^2-7)\phi^{\frac{7}{3}} \\
&\quad +504\phi^{\frac{4}{3}} + 150\nu(\nu-3)\phi + 3(15\nu-7)(7-5\nu) \\
\lambda_0 &= (5\nu-7)(7+5\nu)\phi^{\frac{10}{3}} - 25(\nu^2-7)\phi^{\frac{7}{3}} \\
&\quad -252\phi^{\frac{4}{3}} - 25(\nu^2-7)\phi - (7+5\nu)(7-5\nu)
\end{aligned}
$$

Interested readers are referred to [FEN 98] for more details. It is worth noting that this is the so-called effective porosity (i.e. the porous space filled by the fluid phase and the carbonate crystals), which is the quantity affecting the mechanical moduli. The reason is that the crystals are assumed to transmit only hydrostatic pressure to the cement matrix. With such an assumption, the apparent bulk modulus of the cement matrix saturated by calcite crystals would be that of a saturated porous medium under undrained conditions, the shear modulus being non-affected by the presence of these crystals. This point is further discussed in the section presenting the simulation results.

Concerning the transport properties of the cement matrix, according to [GHA 08] and for such cements, the permeability and the diffusivity coefficients depend on fluid porosity ϕ_F that changes according to[10]:

$$\kappa = 1.2\left(\frac{\phi_F}{0.26}\right)^{11} \cdot \kappa_0 \qquad [\mathrm{m}^2] \qquad [11.26]$$

$$d_{\text{eff}} = 100\exp(9.95\phi_F - 29.08) \qquad [\mathrm{m}^2/\mathrm{s}] \qquad [11.27]$$

Finally, the elastic energy W stored within the solid matrix can be derived easily, thanks to the linearity of the poroelastic equations [11.22], and can be written as follows:

$$
\begin{aligned}
W &= \boldsymbol{\sigma_0} : \epsilon_0 + \frac{\left(\sigma - \sigma_0 + \sum\limits_{J=F,C} b_J(p_J - p_{J,0})\right)^2}{2K} \\
&\quad + \sum_{I,J=F,C} \frac{(p_I - p_{I,0})(p_J - p_{J,0})}{2N_{IJ}} + \frac{(\boldsymbol{s} - \boldsymbol{s_0}) : (\boldsymbol{s} - \boldsymbol{s_0})}{4G} \qquad [11.28]
\end{aligned}
$$

where $\sigma = 1/3\,\mathrm{tr}(\boldsymbol{\sigma})$ denotes the mean stress and $\boldsymbol{s} = \boldsymbol{\sigma} - \sigma\mathbf{1}$ the deviatoric stress.

10 These empirical relations arise from mechanical tests affecting the porosity of the cement matrix. Authors reckon that they may not apply to changes of the porous network due to chemical reactions. Dedicated experimental tests are required to confirm or not the trends given by these relations.

11.2.4. *Balance equations*

In order to complete the description of the problem, the balance equations for the mechanical, hydraulic and chemical subproblems have to be expressed.

The mechanical equilibrium is simply expressed, in the absence of body forces other than gravity and inertia effects, as:

$$\mathbf{div}(\boldsymbol{\sigma}) + \rho \boldsymbol{g} = \mathbf{0} \qquad [11.29]$$

where ρ is the density of the porous medium and $\boldsymbol{g}$ is the acceleration of gravity.

The mass conservation of the in-pore fluid is written in the following form:

$$\frac{1}{\rho_F}\frac{d(\phi_F \rho_F)}{dt} = \frac{1}{\rho_F}\sum_{i=\alpha,w} \mathcal{M}_i \frac{dn_i}{dt} \qquad [11.30]$$

where $\mathcal{M}_i$ is the molar mass of species i. By introducing the bulk modulus of the in-pore fluid, K_F, the variation of the density of the fluid in isothermal conditions can be written in the following form:

$$\frac{1}{\rho_F}\frac{d\rho_F}{dt} = \frac{1}{K_F}\frac{dp_F}{dt} + \sum_{i=\alpha,w} \gamma_i^F \frac{dn_i^F}{dt} \quad \text{with } \gamma_i^F = \left[\rho_F \frac{d(1/\rho_F)}{dn_i}\right]_{p_F, n_{j,j\neq i}} \qquad [11.31]$$

where γ_i^F is the variation of the density of the fluid due to changes in its chemical composition. Moreover, with the assumption of no transmission of shear between calcite and cement matrix, the mechanical behavior of the crystal can be written as:

$$\frac{1}{\rho_C}\frac{d\rho_C}{dt} = \frac{1}{K_C}\frac{dp_C}{dt} \qquad [11.32]$$

where K_C is the bulk modulus of the carbonate crystal, ρ_C its density, and p_C its pressure. Thanks to equations [11.17], [11.18], [11.30] and [11.31] and by assuming a dilute in-pore fluid (i.e. water is in excess as compared to dissolved species), the conservation of fluid mass can be expressed as follows:

$$\frac{\phi_F}{K_F}\frac{dp_F}{dt} + \frac{d\varphi_F}{dt} = (\nu_C^S - \nu_{C,\text{dis}}^F)\mathring{n}_C^S - \sum_{M_i}(\nu_{M_i}^S - \nu_{M_i,\text{dis}}^F \mathring{n}_{M_i}^S) + \text{div}\left(\frac{\kappa}{\eta_{\text{vis}}}\mathbf{grad} p_F\right) \qquad [11.33]$$

According to the assumptions related to reaction kinetics previously introduced, the advancement rate $\mathring{\xi}_{R_i}$ of each reaction R_i can be estimated directly from the quantity of CO_2 supplied to the system by diffusion and advection:

$$\mathring{\xi}_{R_i} = \begin{cases} \mathrm{div}\left(d_{\mathrm{eff}}\,\mathbf{grad}\left(\frac{n_{CO_2}}{\phi_F}\right) + \left(\frac{n_{CO_2}}{\phi_F}\right)\frac{\kappa}{\eta_{\mathrm{vis}}}\mathbf{grad}p_F\right) \\ \quad \text{if } M_{1,R_i} \neq 0 \text{ and } M_{1,R_{i-1}} = 0 \\ 0 \text{ otherwise} \end{cases} \quad [11.34]$$

The variation of the molar quantity of CO_2 corresponds to the gas quantity brought by advection and diffusion minus the quantity consumed by chemical reactions:

$$\frac{dn_{CO_2}}{dt} = \mathrm{div}\left(d_{\mathrm{eff}}\,\mathbf{grad}\left(\frac{n_{CO_2}}{\phi_F}\right) + \left(\frac{n_{CO_2}}{\phi_F}\right)\frac{\kappa}{\eta_{\mathrm{vis}}}\mathbf{grad}p_F\right) - \sum_{R_i}\mathring{\xi}_{R_i} \quad [11.35]$$

Finally, equations [11.29] and [11.33]–[11.35] define the chemo-poromechanical model that has to be solved.

11.3. Application to wellbore cement

11.3.1. *Description of the problem*

The coupled chemo-poromechanical problem described above has been implemented into a numerical code to study the effects of the presence of dissolved CO_2 in contact with the cement constituting the injection well in a CO_2 geological storage complex. In a first approach, the problem geometry has been simplified. We chose to work in an axial-symmetry configuration, the axis being the well's axis. Furthermore, assuming that the region of interest is far from boundary effects such as the injection point, geological faults, or the well's head, we also assume to work in plane strain. These two simplifications imply that all the quantities involved in the problem only depend on the radial coordinate (distance to the well's axis). The problem is thus reduced to a 1D-problem.

This problem has been implemented into a numerical code using a finite difference scheme for the time discretization and the finite volume method for the spatial discretization. This latter has been chosen for its good properties when dealing with reactive transport problems (conservative resolution for the mass balance equation for instance or numerical stability in the presence of sharp fronts). One of the key aspects of the resolution strategy adopted here is that the problem (equations [11.29] and [11.33]–[11.35]) is solved in a fully coupled way.

11.3.2. *Initial state and boundary conditions*

We consider a hollow-cylindrical domain made of cement with an internal radius of 89 mm, an external radius of 108 mm and an unit height. CO_2 will be assumed to reach the cement through its external surface, thus simulating a leakage of dissolved gas in water through the interface between the caprock and the well. We consider a uniform fluid pressure of 13 MPa at the beginning, which roughly corresponds to the injection pressure of CO_2 at a depth approximately 1,000 m. The *in situ* stress field is uniform and isotropic. The mean total stress is 25 MPa. In addition, the cement is strong enough and without any carbonate crystal before the injection of CO_2. The initial porosity is 30%, a realistic value encountered in class G cements. The in-pore water does not contain dissolved CO_2 at the initial state.

At the beginning of the injection stage ($t = 0$ s), the fluid pressure of 13 MPa is kept constant on the outside face of the cylinder and, with the arrival of CO_2-rich fluid, a CO_2 concentration corresponding to the saturated concentration of CO_2 in water is fixed on this surface. On the inside face, because of the presence of the metallic casing, an impermeable boundary is assumed. Finally, concerning the mechanical subproblem, we choose to apply a radial stress of compression of 25 MPa on the external surface and of 13 MPa on the internal surface of the well, which corresponds to the stress induced by injection pressure.

11.3.3. *Illustrative results*

The cement domain has been exposed to the presence of CO_2 at its outer wall (well–caprock interface). The simulated period cover 180 days, which is sufficient for the complete carbonation of the cement sheath. The material parameters used in the simulations presented hereafter are summarized in Table 11.1.

ϕ_0 (–)	κ_0 (–)	η_{vis} (m^2/s)	K_m (MPa)	G_m (MPa)	K_F (MPa)
0.30	10^{-19}	$1.79 \cdot 10^{-9}$	17,500	10,575	2,200

ν^S_{CH} (cm^3/mol)	$\nu^S_{CSH1.6}$ (cm^3/mol)	$\nu^S_{CSH1.2}$ (cm^3/mol)	$\nu^S_{CSH0.8}$ (cm^3/mol)	$\nu^S_{SiO_2}$ (cm^3/mol)	ν^S_{C} (cm^3/mol)
33.1	84.7	72	59.3	29	36.9

Table 11.1. *Material parameters*

Figure 11.3 illustrates the evolution of the chemical reactions as a function of space at several times. It gives the evolution of the molar quantities of CH, C–S–H and $CaCO_3$. We recall that the cement is exposed to CO_2 from its outer surface. A clear carbonation front can be observed in Figure 11.3(a). After 30 days of exposure, the domain comprised between $r = 100$ mm and $r = 108$ mm is fully carbonated and does not contain CH and C–S–H anymore. Figure 11.3(b) shows that the cement domain is fully carbonated after 161 days of exposure to CO_2. After this time, all the CH and C–S–H have been consumed, leaving room for the carbonate crystals. As seen previously, these changes in the composition of the cement matrix involve evolutions of its porosity. Figure 11.4(a) shows how the chemical reactions affect the fluid porosity ϕ_F.

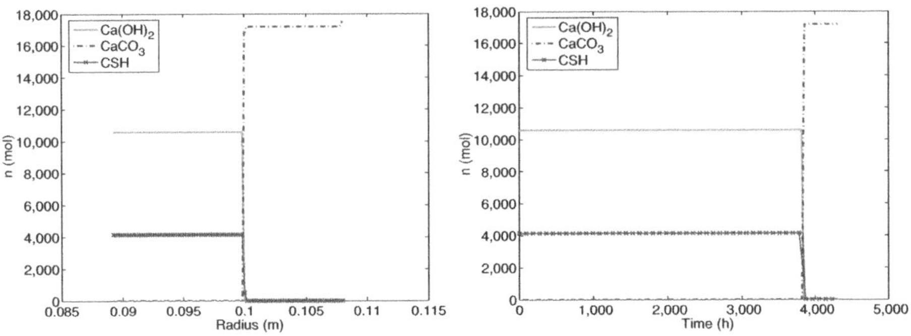

(a) Spatial evolution after 30 days of exposure. (b) Time evolution at well inner wall ($r = r_{int}$).

Figure 11.3. *Evolution of the molar quantities of CH, C—S—H, and $CaCO_3$ due to the presence of CO_2 at the well–caprock interface*

When the cement sample is strong (i.e. before being carbonated), its porosity is 30%, whereas in the part completely carbonated, its porosity decreases down to 24.4%. This loss of porosity involves, as seen from equations [11.26] and [11.27], an alteration of the transport properties such as a decrease in the permeability, which is divided by 10 and of the diffusivity coefficient of the CO_2 in water, which is divided by 2 in the carbonated zones.

Moreover, as a result of the production of water (see chemical reactions R_1 to R_4) and decrease in the fluid porosity due to the formation of the carbonate crystals, a built-up of the in-pore fluid pressure is observed. Owing to the relatively low permeability of the cement in its strong state and its further reduction because of the forming crystals, this pressure built-up cannot be immediately evacuated. This behavior can be observed in Figure 11.4(c) where a peak of the in-pore fluid pressure localized at the carbonation front appears. This peak induces a water flow directed toward the external surface of the cement sheath.

Even though the fluid porosity decreases with the carbonation, the dissolution of the cement matrix due to its leaching leads to an increase in the effective porosity of the matrix ϕ (see Figure 11.4(b)). This porosity of the cement matrix (excluding the carbonates) varies between 30% and 82% in the simulation. Consequently and according to equation [11.24], the cement matrix stiffness is highly degraded: the bulk modulus is divided by 9 and the shear modulus by 3 in the carbonated zones. This indicates a potential risk for the cement. However, it must be recognized that the calcite will certainly precipitate in pores that would have appeared within the cement matrix because of the carbonation of its hydrated phases. As a result, the precipitated calcite filling these pores would probably stiffen the material (particularly in terms of shear modulus) and should probably be taken into account in the estimation of the overall stiffness of the cement.

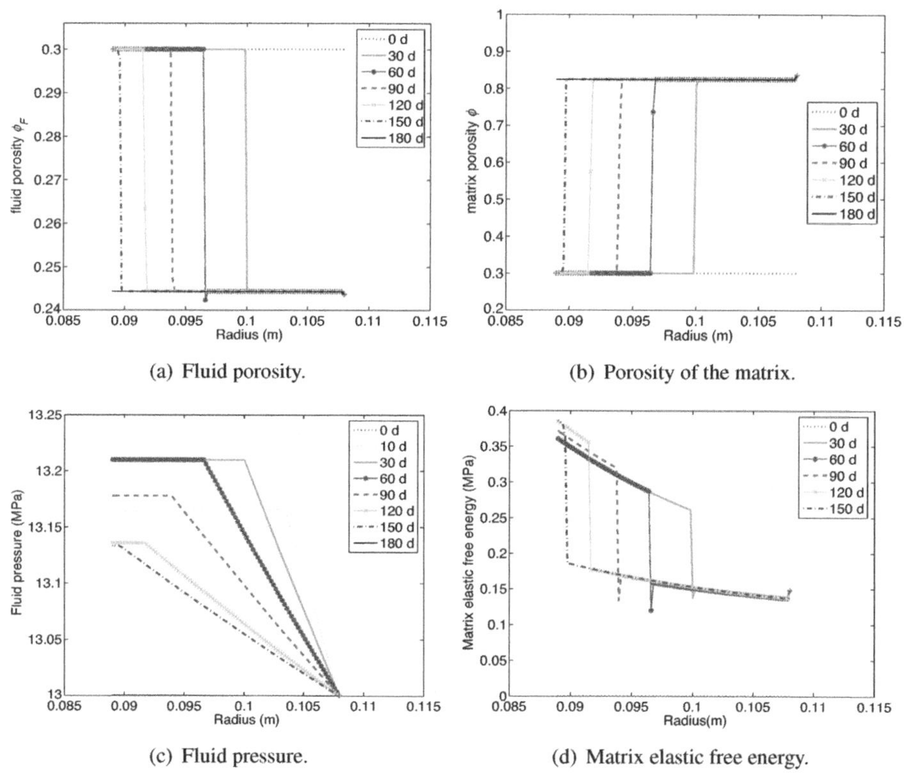

(a) Fluid porosity.

(b) Porosity of the matrix.

(c) Fluid pressure.

(d) Matrix elastic free energy.

Figure 11.4. *Evolution of porosities, fluid pressure, and matrix elastic free energy in the cement sample during 180 days of exposure to aqueous CO_2*

The risk of damage of the cement can be evaluated by the estimation of the elastic free energy stored in the cement matrix. As seen in Figure 11.4(d), the elastic free energy increases because of the chemical reactions. This could lead to some damage. It is interesting to note that this increase in the free energy is higher when the initial permeability decreases or when the diffusion coefficient increases. In the first case, the water produced by the carbonation process cannot be evacuated from the cement sheath so easily, and in the second case, the water is produced more rapidly. In both the cases, the peak of pressure is higher. As an illustration, Figures 11.6(a) and 11.6(b) show the elastic free energy of the matrix for $\kappa = 6 \cdot 10^{-22}$ m^2 and for $\kappa = 6 \cdot 10^{-21}$ m^2. The value of this pressure peak thus evolves in a range comprising several orders of magnitude. Further analyses are required to quantify the potential damage of the cement induced by this peak. Recent studies have shown that well cement may present damage signs when carbonated in aqueous conditions representative of CO_2 geological storage conditions (see for instance [FAB 09]). Besides the fluid pressure peak, the crystallization of calcite could also lead to local stress concentrations on the cement matrix, at the carbonation front in particular. However, these effects have not been accounted for in the present study.

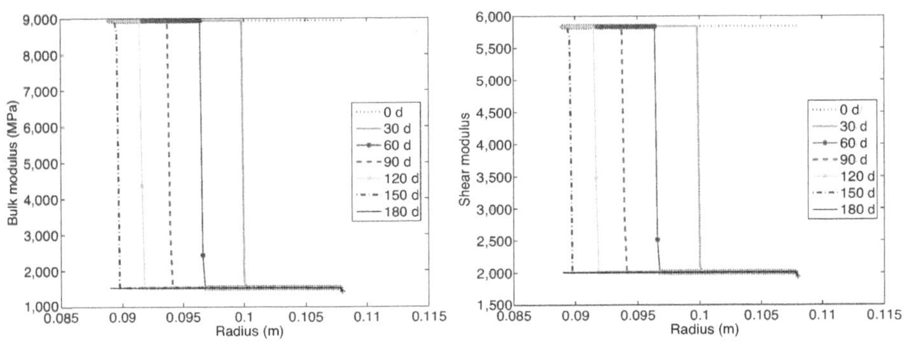

(a) Bulk modulus (b) Shear modulus

Figure 11.5. *Evolution of the mechanical moduli of the cement during 180 days of exposure to aqueous CO_2*

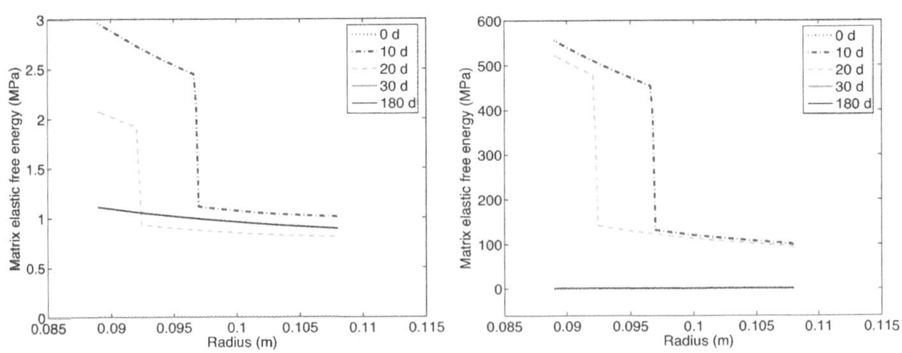

(a) $\kappa = 6 \cdot 10^{-21}$ m^2 (b) $\kappa = 6 \cdot 10^{-22}$ m^2

Figure 11.6. *Evolution of the matrix elastic free energy in the cement sheath during 180 days of exposure to aqueous CO_2: effect of the initial permeability. a) $\kappa = 6 \cdot 10^{-21}$ m^2 and b) $\kappa = 6 \cdot 10^{-22}$ m^2*

11.4. Conclusion

A general framework to model chemo-poromechanical problems has been presented and discussed. The case of cement carbonation in the context of CO_2 geological storage has been used to illustrate the proposed methodology. It is believed that this work may easily be extended or transposed to other geomechanical problems involving precipitation/dissolution reactions.

The simulation of the carbonation of the injection well when subjected to the presence of aqueous CO_2 has been presented. It shows that under certain circumstances, a peak of fluid pressure can appear and may be relatively large. This peak could lead to structural damage. It has also been shown that the occurrence of this pressure peak can be avoided if the cement formulation is appropriate (relatively high permeability for instance).

11.5. Acknowledgments

The authors would like to acknowledge Dr. Antonin Fabbri and Dr. Henry Wong from ENTPE (Lyon, France) for their insightful discussion on this topic.

11.6. Bibliography

[COU 02] COUSSY O., DANGLA P., "Approche énergétique du comportement des sols non saturés", in COUSSY O., FLEUREAU J.-M., (eds), *Mécanique des sols non saturés*, Hermes-Lavoisier, Paris, pp. 137–174, 2002.

[COU 04] COUSSY O., *Poromechanics*, John Wiley & Sons, Chichester, 2004.

[COU 05] COUSSY O., "Poromechanics of freezing materials", *Journal of the Mechanics and Physics of Solids*, vol. 53, no. 8, pp. 1689–1718, 2005.

[COU 10] COUSSY O., PEREIRA J.M., VAUNAT J., "Revisiting the thermodynamics of hardening plasticity for unsaturated soils", *Computers and Geotechnics*, vol. 37, no. 1–2, pp. 207–215, 2010.

[DAN 02] DANGLA P., "Plasticité et hystérésis", in COUSSY O., FLEUREAU J.-M., (eds), *Mécanique des sols non saturés*, Hermes-Lavoisier, Paris, 2002.

[FAB 09] FABBRI A., CORVISIER J., SCHUBNEL A., BRUNET F., GOFFÉ B., RIMMELE G., BARLET-GOUÉDARD V., "Effect of carbonation on the hydro-mechanical properties of Portland cements", *Cement and Concrete Research*, vol. 39, pp. 1156–1163, 2009.

[FEN 98] FEN-CHONG T., Analyse micromécanique des variations dimensionnelles de matériaux alvéolaires – Application au polystyrène expansé, PhD Thesis, Ecole Polytechnique, Paris, 1998.

[GHA 08] GHABEZLOO S., Comportement thermo-poro-mécanique d'un ciment pétrolier, PhD Thesis, Ecole Nationale des Ponts et Chaussées, 2008.

[HOU 97] HOULSBY G.T., "The work input to an unsaturated granular material", *Géotechnique*, vol. 47, no. 1, pp. 193–196, 1997.

[MAI 99] MAINGUY M., Modèles de diffusion non linéaires en milieux poreux. Application à la dissolution et au séchage des matériaux cimentaires, PhD Thesis, Ecole Nationale des Ponts et Chaussées, Paris, 1999.

[PER 05] PEREIRA J.M., WONG H., DUBUJET P., DANGLA P., "Adaptation of existing behaviour models to unsaturated states: application to CJS model", *International Journal for Numerical and Analytical Methods in Geomechanics*, vol. 29, no. 11, pp. 1127–1155, 2005.

[THI 05] THIERY M., Modélisation de la carbonatation atmosphérique des bétons, PhD Thesis, Ecole Nationale des Ponts et Chaussées, Paris, 2005.

[VAN 10] VANDAMME M., BROCHARD L., LECAMPION B., COUSSY O., "Adsorption and strain: the CO_2-induced swelling of coal", *Journal of the Mechanics and Physics of Solids*, vol. 58, no. 10, pp. 1489–1505, 2010.

[XU 06] XU T., SONNENTHAL E., SPYCHER N., PRUESS K., "TOUGHREACT–A simulation program for non-isothermal multiphase reactive geochemical transport in variably saturated geologic media: applications to geothermal injectivity and CO_2 geological sequestration", *Computers & Geosciences*, vol. 32, no. 2, pp. 145–165, 2006.

List of Authors

Gérard BALLIVY
University of Sherbrooke,
Canada

Laurent BROCHARD
Ecole des PontsParisTech – IFSTTAR – CNRS
France

Daniel BROSETA
University of Pau and Pays de l'Adour
France

Nicolas BURLION
University of Lille 1
France

Robert CHARLIER
University of Liège
Belgium

Frédéric COLLIN
University of Liège
Belgium

Patrick DANGLA
Ecole des PontsParisTech – IFSTTAR – CNRS
France

Rémi DE LA VAISSIÈRE
ANDRA
Agence nationale pour la gestion des déchets radioactifs
France

Teddy FEN-CHONG
Ecole des PontsParisTech – IFSTTAR – CNRS
France

Jian FU SHAO
University of Lille 1
France

Adeline GARNIER
TOTAL Exploration-Production
France

Pierre GÉRARD
University of Liège
Belgium

Sylvie GRANET
EDF R&D
France

David GRÉGOIRE
University of Pau and Pays de l'Adour
France

Nicolas GUY
BRGM & ENS Cachan
France

François HILD
ENS Cachan/CNRS/UPMC/PRES UniverSud
France

Y. JIA
University of Lille 1
France

Fadi KHADDOUR
University of Pau and Pays de l'Adour
France

Christian LA BORDERIE
University of Pau and Pays de l'Adour
France

Brice LECAMPION
Schlumberger SA
USA

Jolanta LEWANDOWSKA
University of Montpellier 2
France

Saeid NIKOOSOKHAN
Ecole des PontsParisTech – IFSTTAR – CNRS
France

Olivier NOUAILLETAS
University of Pau and Pays de l'Adour
France

Roland J.-M. PELLENQ
MIT & CNRS
USA

Jean-Michel PEREIRA
Ecole des PontsParisTech – IFSTTAR – CNRS
France

Céline PERLOT
University of Pau and Pays de l'Adour
France

Gilles PIJAUDIER-CABOT
University of Pau and Pays de l'Adour
France

Jean-Pol RADU
University of Liège and F.R.S.-F.N.R.S.
Belgium

Baptiste ROUSSEAU
University of Sherbrooke,
Canada

Jeremy SAINT-MARC
TOTAL Exploration-Production
France

A.P.S. SELVADURAI
McGill University
Canada

Darius M. SEYEDI
BRGM
France

Jitun SHEN
Ecole des PontsParisTech – IFSTTAR – CNRS
France

Serigne SY
BRGM
France

Jean TALANDIER
ANDRA
Agence nationale pour la gestion des déchets radioactifs
France

Mickaël THIERY
IFSTTAR
France

Valérie VALLIN
Ecole des PontsParisTech – IFSTTAR – CNRS
France

Matthieu VANDAMME
Ecole des PontsParisTech – IFSTTAR – CNRS
France

Index

Printed and bound by CPI Group (UK) Ltd, Croydon, CR0 4YY

13/09/2026

14961075-0003